A Film-Philosophy of Ecology and Enlightenment

Inspired by the philosophy of Wittgenstein and his idea that the purpose of real philosophical thinking is not to discover something new, but to show in a strikingly different light what is already there, this book provides philosophical readings of a number of "arthouse" and Hollywood films, connected by the theme of trauma and recovery, recovery in the form of awakening. Each chapter contains a discussion of two films—one explored in greater detail and the other analyzed as (and in) a minor key which reveals the possibility for the book's ideas to be applied across different films, registers and genres. The readings are not only interpretive, but they offer a way of thinking and feeling *about*, *with* and *through* films that is genuinely transformative. Rupert Read's main contention is that certain films can bring about a change in how we see the world. He advocates an ecological approach to film-philosophy analysis, arguing that film can re-shape the viewer's relationship to the environment and other living beings. The transformative "wake-up call" of these films is enlightenment in its true sense. The result is a book that ambitiously aims to change, through film, how we think of ourselves and our place in the world, at a time when such change is more needed than ever before.

Rupert Read is Reader in Philosophy at the University of East Anglia, UK. He is a renowned Wittgensteinian scholar, with major research interests in political and environmental philosophy. His published monographs to date are *Kuhn* (co-authored, 2002), *Applying Wittgenstein* (2007), *Philosophy for Life* (2007), *There is No Such Thing as a Social Science* (2008), *Wittgenstein Among the Sciences* (2012), and *A Wittgensteinian Way with Paradoxes* (2012). His editorial experience includes *The New Hume Debate* (co-edited, 2000), *Film as Philosophy: Essays on Cinema after Wittgenstein and Cavell* (2005), and the work for which he is perhaps still best known, *The New Wittgenstein* (Routledge, 2000), which offers a major re-evaluation of Wittgenstein's thinking.

Routledge Research in Aesthetics

Michael Fried and Philosophy
Modernism, Intention, and Theatricality
Edited by Matthew Abbott

The Aesthetics of Videogames
Edited by Jon Robson and Grant Tavinor

Tragedy and Redress in Western Literature
A Philosophical Perspective
Richard Gaskin

The Pleasure of Pictures
Pictorial Experience and Aesthetic Appreciation
Edited by Jérôme Pelletier and Alberto Voltolini

Thinking with Images
An Enactivist Aesthetics
John M Carvalho

A Film-Philosophy of Ecology and Enlightment
Rupert Read

For more information about this series, please visit: www.routledge.com/Routledge-Research-in-Aesthetics/book-series/RRA

A Film-Philosophy of Ecology and Enlightenment

Rupert Read

Routledge
Taylor & Francis Group
LONDON AND NEW YORK

First published 2019 by Routledge

2 Park Square, Milton Park, Abingdon, Oxfordshire OX14 4RN
52 Vanderbilt Avenue, New York, NY 10017

Routledge is an imprint of the Taylor & Francis Group, an informa business

First issued in paperback 2020

Library of Congress Cataloging-in-Publication Data
A catalog record for this book has been requested

ISBN: 978-1-138-59602-3 (hbk)
ISBN: 978-0-367-66491-6 (pbk)

Typeset in Sabon
by Apex CoVantage, LLC

See demons as demons: that is the danger.
Know that they are powerless: that is the way.
Understand them for what they are: that is deliverance.
Recognize them as your father and mother: that is their end.
Realize that they are creations of the mind: they become its glory.
When these truths are known, all is liberation.

—Milarepa

Contents

Preface

Film is the great art form of our time. I shall set out in the Introduction to the present work some of the ways I see film as contributing—in specific fashions that other art forms or media sometimes cannot—to the great questions of our time: by way of "point of view," by way of its peculiar and grand possibilities for both immersion (e.g. 3-D) and "alienation" (*á la* Brecht).

In this Preface, I want to provide the merest indication of the "backstory" to why this book (therefore) focuses on the art form of film—from the ultimate in arthouse (such as *Last Year in Marienbad*) to the ultimate in big box office (such as *Avatar*),[1] and to indicate what pre-conditions there may be to approaching these films in the way I do.

In 2005, my book (co-edited with Jerry Goodenough) entitled *Film as Philosophy: Essays on Cinema after Wittgenstein and Cavell* was published. This was the first book to bring together the main voices advocating that films can genuinely function as philosophical texts, and exemplifying that claim across a series of impressive cases of philosophical films.

Especially since then, there has been a great deal of interest in the question of whether films can function as philosophical works. This interest however seems sooner or later inevitably to founder on the following dilemma: either the philosophical work done by films is paraphrasable, in which case ultimately the films in question are merely pretty or striking *vehicles* for philosophizing which precedes them; or the philosophical work done by films is not paraphrasable, in which case it seems mysterious / dubious / systemically obscure.

However, this dilemma, while in its own terms quite correct, rests, I submit, on an unjustified presumption. The presumption is that philosophical "work" has to be understood (if it is to be worthwhile) as issuing in views/opinions/theses/theories (the *content* of the would-be paraphrase). But there is another possibility, a possibility explored at greatest length in Wittgenstein's philosophy: that philosophical work at its best is "therapeutic," in very roughly the psychoanalytic sense of that word.[2] Or better still, that philosophical work is "liberatory": essentially freeing us from unaware constraint by views.[3] (Views not in the sense of viewings, seeings or ways of seeing, as with viewings of films or possible perspectives; these are great. Views rather in the standard philosophical sense simply of opinions, or would-be definitive claims, most usually, claims of *essence*, claims of *necessity*.)[4] Philosophy need not—and in fact, if one is at all taken by Wittgenstein's quiet philosophical revolution, should not—issue in any controversial theses or opinions, any theories, at all. Rather, it should work with the "patient's"—the interlocutor's, the co-conversationalist's, the other's, and indeed one's own—presumptions, exposing them to awareness, and thus empowering her/him to autonomously acknowledge, justify, or overcome or transform them, where necessary.

My 2005 *Film as Philosophy* collection, and especially Hutchinson's and my essay in that collection, endeavoured in a preliminary way to develop the idea sketched above. In the present book, I enter significantly further into it, and into the following closely associated question: Is there a way to understand how some of the greatest (including *popular*) films work that transcends any heresies of alleged paraphrasability,[5] transcends theories that would subject films to their diktat, and empowers the viewer to understand the work that the films in question do as liberatory work upon and *with* and through them, the viewer? A difficulty seemingly facing the efforts to understand films as philosophical works has been (in most cases) their consistently "dialogical" nature, the way that they offer different voices, and not just (as most philosophical prose works do) one voice. However, this is a *strength* of these film-as-philosophical works once they are understood as "liberatory" works. Thus this book does not focus primarily let alone exclusively on "arthouse" films (though it includes plenty of these). It focuses in the end more on films liable to have a wider influence, films which in that sense matter more. Because I dare to venture (in the chapters on *Gravity* and *2001*, and on *Lord of the Rings* and *Avatar*) that some films are popular *because they are good*. Because they are mythic or neo-mythic. *Because they are rich and "dialogical" and thus surprisingly open to being experienced as philosophy*.

It was, obviously, not an accident of my 2005 collection that its subtitle referenced Wittgenstein (and Cavell). For it is his (their) thought that has (above all) inspired most of those who I take to be the best contemporary writers on film as philosophy. They take up a very different stance from that found in most film studies and film theory. A stance that, crucially, opens to the films, and allows them space to breathe (rather than dictating to them via spectating on them from a position of "superiority").

Now I have written a book I consider to be pursuing much the same approach in detail, across a series of what I think are some of the very best philosophical films ever made. My philosophical approach is deeply inspired by Wittgenstein. This book maintains an openly Wittgensteinian, anti-theoretical, anti-elitist stance, and yet a stance that does not hesitate to engender judgements of aesthetic quality. This manifests also in my selection of material, which (as already noted) prominently includes both high-arthouse and certain mega-blockbuster films, side by side.[6] (It is also manifested in the way in which) I take these films to manifest a kind of call to action. A call which necessarily goes beyond the academic.

A reader of the present work versed in philosophy will profit from being so versed. Still more so, if that philosophy is Wittgensteinian: for Wittgenstein's influence is present right across the piece of this book, even where his name is absent (as, mostly, it is). *But* (and partly through reserving some more heavily philosophical material to the endnotes) I have sought to design the book in such a way that readers who lack prior philosophical training or acquaintance with Wittgenstein will nevertheless have little difficulty in parsing the text, in reading the "readings" of films offered here. (However, I DO assume prior familiarity with the *films*.)

If you are such a philosophically innocent reader, and if you meet a part of the text that philosophically baffles or eludes you, my advice is simple: try to understand it; and if you really cannot, then just skip it, confident in the understanding that the essence of what is important in the present work is contained ultimately in the films themselves, and that you can come to understand *that* by understanding the heart or spirit of what I've written, regardless of prior philosophical orientation or education.

Artistic representations, if they are good enough and powerful enough, can catalyse. They can impact our worldview, by changing how something is conceptualized and presenting it in the context of an actual life, which abstract thinking on its own cannot do. The very future of our living planet is linked to the possibility of evolving attitudes towards our place in it. Film, as the great mass medium of our time, may turn out to have a vital part to play, if there is to be a future for us. This book seeks to indicate some of the dimensions of that part, that task.

And to begin to pursue it.

***

Thanks to all those who helped enable this book. Those who helped with individual chapters are acknowledged in them. But I want particularly to mention here those who have been vital across the field of the whole book:

My teachers, Stephen Mulhall and Stanley Cavell.
The good people at Routledge.
My students in my pathfinding "Film as Philosophy" classes, over the last 20 years at UEA.
Those who have taught film as philosophy alongside me: Emma Bell, Vincent Gaine and Jerry Goodenough.
The ThinkingFilm collective, including Phil Hutchinson.
Participants in my Merton Road film "salon."
My colleague in working on *Avatar* and much more besides, Peter Kramer.
And finally Silvia Panizza, for important help in conceptualizing the book manuscript (and for some editorial assistance).

The chapter on *Avatar* includes material first published in my "The Call of *Avatar*," in 2010, in issue 4 of *Radical Anthropology* (http://radicalanthro pologygroup.org/sites/default/files/journal/journal_04.pdf); this is reproduced by kind permission.

The chapter on *Melancholia* includes material first published in my "An Allegory of a 'Therapeutic' Reading of a Film: Of MELANCHOLIA," in 2014, in SEQUENCE; this is reproduced by kind permission.

The chapter on *Gravity* includes material first published in "Gravity's Pull," in 2014, on the ThinkingFilm website: http://thinkingfilmcollective.blogspot.com/2014/01/gravitys-pull.html. Thanks to Peter Kramer, my co-author, for permission to rework this material in my own name.

Notes

1. This ecumenism is one of the main ways in which my approach differs from the (otherwise like-minded) approach found in the concept of "the essay film." Another is that my approach, as will be described below, tends to be Wittgensteinian, and thus more hostile to "theory" than most (though by no means all) advocates of "the essay film" are. Another is that I am in any case less interested in auteurs, more (following the old "New Critics") in the artwork as produced. (Though that is not to deny that I think most of the directors whose films feature in this book geniuses. But quite often, when I use their names in this book, it is really shorthand for the necessarily *ensemble*-like character, the collective intentionality, manifested in virtually all large-scale filmic artwork.)
2. To understand the sense of the term "therapy" that is in play here, see my and Crary's The New Wittgenstein (London: Routledge, 2000), and the later work of Gordon Baker (especially his Wittgenstein's Method: Neglected Aspects (Oxford: Blackwell).
3. See my Liberatory Philosophy, forthcoming, for detail.
4. On this point, see Oskari Kuusela's work, especially his The Struggle Against Dogmatism (Cambridge, MA: Harvard, 2008).
5. See Cleanth Brooks, "The heresy of paraphrase," in his The Well-Wrought Urn (New York: Harvourt, 1975 (1947), the greatest of the masterpieces of New Criticism. Brooks makes clear how something that is fully paraphrasable is not art.
6. Obviously, there are significant precedents for this: not least, in the work of those who were my teachers in the philosophy of film, Stephen Mulhall and the late Stanley Cavell! And obviously, it is commonplace now to consider the philosophy of everything from *Mission: Impossible* to *The Simpsons*. But I would venture the claim that the popular films I treat of here are treated more respectfully (in the sense that I seek to avoid projecting ideas onto them), *and yet* without exaggeration (e.g. I think that some of the claims Mulhall makes, in the second edition of On Film (London: Routledge, 2001), for the *Mission: Impossible* films are a bridge too far), than in most of those "philosophy of" pieces/books.

Introduction

Film as Freedom: The Meaning of Film as Philosophy

> We use our imagination not to escape the world but to join it, and this exhilarates us because of the distance between our ordinary dulled consciousness and an apprehension of the real.
>
> —Iris Murdoch, *The Sovereignty of Good*

> Myth is not entertainment, but rather the crystallization of experience, and far from being escapist literature, fantasy is an intensification of reality.
>
> —Alan Garner, quoted in David Loy, *The World Is Made of Stories*

This book considers films as philosophical investigations. And submits that these investigations both constitute and conduce towards (both) true intellectual freedom *and* ecological wisdom. And that these two are in the end two sides of the same coin.

The main contention that animates the book is that certain films (of which those analysed in the book are among the finest examples) can occasion our bringing about a needful change in how we see the world, from which a change in lifestyle and action ensues. The films however are not just simply paraphrasable "messages" in disguise; they are invitations to a "dialogue" and to profound *reflection*.

An original feature of this book is its *active* concern with how some films can and do and should and will change our relationship with "the environment" (better: with the other beings that together with us co-constitute the living world). These films act as radical eye-openers about the role of humans in the world, and thus help us *see* ourselves as part of the living ecological system. These films, in short, free us from (dangerous) widespread contemporary prejudices and delusions about our alleged freedom from ecological "constraints"; from reality itself.[1]

Inspired by the philosophy of Wittgenstein and his idea that what real philosophical thinking does is not to discover something new, but to show in a strikingly different light what is already there, I aim to do just that. To show what is in these films, and thus in our world. As it is. And as it might be. For ill—or for good.

The goal is[2] liberation from attachment to (compulsion by) false, arbitrary or misleading pictures and a newly acquired ability to think and perceive more clearly and honestly, less denialistically, and thus (ultimately) more ecologically.

The method of the book, as a consequence, differs from that of most philosophical or theoretical work on film. Rather than using theory as a pre-formed tool with which to understand the films, I allow philosophical, political and ecological ideas to stand alongside the films and *to emerge from the films themselves*. Nor does this book "find" Wittgenstein's philosophy allegedly hidden in the films; rather, *it finds the films*. In other words, Wittgenstein helps *orient* one towards the films so that one can (as Wittgenstein famously puts it)[3] "find one's way about" both in them and in relation to them. While the overall approach of the book is based on the idea that changes in thought are also changes in perception and hence already commence changes in response and action, the specific thoughts (and related perceptions and actions) of course vary somewhat from film to film.

Often, it takes an "external" stimulus to jolt us into awareness of an aspect of a situation or of a problem, such as a prejudice we may have (concerning our situation). Films provide such stimuli, and the more they do it in effective *and* yet subtle and complex ways, the better their artistic value. Hence, the criterion according to which the films have been selected for the book is their depth and filmic quality and applicability. In this respect, too, my book makes a still slightly controversial contribution by focusing on both "arthouse" *and* mass-market films without differentiation, for some of the latter can create just such intelligent awareness. Good films, from the perspective of the present work, are simply those that, when understood aright, can effect a thoughtful change in the (thoughtful) viewer.

Many of the films featured in this book have of course been much discussed before. But the films haven't been brought together in the way they are here. Nor, in some cases, have they been thought of as *ecological* films (see especially Chapters 1, 2 and 5), or as films capable of genuinely *enlightening* the viewer (see especially Chapters 1, 5 and 6). Few of these films have been thought of film-philosophically, and virtually none of them from a Wittgensteinian perspective. Finally, the way in which this book suggests that ecology and enlightenment are two sides of the same coin, and (so) that there is no contradiction between ecology and freedom, once the latter is understood aright, is new, and is unveiled and explored by means of / *in* these films.

Each chapter in this book contains a discussion of two films, one explored in detail and the other as "minor key" to it.[4] This structure makes it possible for the ideas developed in connection with the first film to expand beyond their original context.

As a would-be "therapy" for our unwell society, the book shows an awareness of the historical development of the culture it addresses, by analyzing films which span the history of post-war cinema, though with a greater focus on relatively contemporary productions. Ambitiously, this book aims at changing, through these film-viewings, how we think of ourselves, our place in the world—and also what we (might or will) do about the grave problems confronting us. How we might yet do so freely (but without getting stuck in versions of freedom that in fact corral us and make clear vision impossible).

Before going further, I should note that there is a general background state of affairs to this book that I take for granted. I shall not *argue* here for the claim that our society is ill; that our situation is very serious, probably desperate; that we are experiencing ecological disaster[5] and the only question is whether we can stop that becoming ecological-and-human catastrophe;[6] that we are mostly not, as we may too easily fantasize ourselves to be, as consumers (and maybe even citizens) in the contemporary West, free. I shall not, that is, in short, spend time here *arguing* that things in our world are significantly worse than the mainstream media virtually ever lets on, than our "left vs. right" politics assumes (stuck as is it is within the frame of economism and endless "growthism"),[7] and than conventional academic discourse typically takes for granted.[8] Others and myself have argued for the claims I here *take for granted*—and have radically questioned the conventional assumptions that are *normally* taken for granted—elsewhere;[9] and this book is not the place to rehearse those arguments (although I shall periodically *reference* them as we go along).

I do however need to say a little more here about how I think we ought to use the vital term, "freedom."

Let us then commence a recasting of what freedom *is*, by means of the kinds of transformative film-works that this book focuses upon. The title of the present work operates with terms that, while apparently in contradiction, can work together to re-shape the idea of freedom away from its traditional Western-liberal sense (broadly, as lack of constraint). Real freedom is *here* understood rather to link "enlightenment" and "ecology" by referring to something that typically we achieve *together* (as in a revolution where an authoritarian government is toppled by civic society; or as in a true democracy; or indeed as in a film we watch together that precipitates the scales falling from our eyes). Including the freedom to assent to and find ourselves at home in what we inescapably are (i.e. part of the broader ecological system). This both brings about and is itself already a new form of enlightenment, of an indissolubly individual *and* collective kind. (This book recovers the sense of "enlightenment" present in Buddhism [discussed in Chapter 6]; in which enlightenment is not opposed to community or ecology, but [centrally] involves foundation in and reconciliation with these. I suggest that what in Europe we call *The* Enlightenment badly needs tempering by this Eastern sense of the concept.) Western culture has long operated with the assumption that we can survive and prosper by means of thinking, seeing and acting anthropocentrically, fixating on humans alone. But, properly understood, anthropocentrism collapses into *ecocentrism*,[10] the placing of ecosystems front and centre: the sort of perspective fairly clearly visible, to the uncaptive eye, in most of the films discussed in this book.

The shift in perspective which this book aims at, and which characterizes Wittgenstein's philosophical goal, is an ethical achievement. If the change in perception is towards greater truthfulness and justice and love, then making it possible becomes itself a moral task. This idea, which underlies the book's own method, connects me to Iris Murdoch, whose moral philosophy revolves around the possibility of attending to reality in order to perceive its moral content (and demands).[11] This is done through what Murdoch calls the "loving gaze," which we could understand here as the desired mode of perception of a responsible film viewer. Seeing films lovingly means loving them in their beauty or grandeur or wisdom, and allowing them to make ethical demands on us, removing the obstacles to (clear) perception that come from prior fears, commitments, or interests. Seeing films lovingly then simultaneously means (as per my epigraphs) gazing (attentively, lovingly) *through* them at the world they disclose. The paradise we inhabit, the greater paradise we could turn it into (and of course, the hell we too often make it and currently threaten to confine it into forever).

And the wonderful thing that helps this task is: film itself encourages attention. I mean not only what is obviously true, that any fine film naturally demands the viewer's attention and absorption. Nor do I mean "just" that some of the films of such directors as (say) Terrence Malick, Orson Welles and Jane Campion are partly *about* or manifestative of joint deep attention—*as are most of the films in this book*. Nor even do I mean only that I am very interested in this book in films, such as *Waltz With Bashir*, *Last Year in Marienbad*, *2001: A Space Odyssey* and *Avatar*—and also such films as *Blade Runner*,[12] *The Prestige*,[13] Tarkovsky's *Mirror*, *Koyaanisqatsi*, *Timecode*, *L'avventura*, *Sunday Bloody Sunday* and some of the films of Peter Greenaway, Michael Haneke and Hitchcock—which are *inter alia* specifically about joint deep attention *in cinema*. I mean, more fundamentally, simply that the setting and the act, the experience of watching something together, in the darkness, with a grand screen before one, is *itself* a way that one is inclined, aided, gently directed towards *presence*. Towards being truly present, with a quasi-meditative attention. (Thus it is important that film remains tied to *cinema(s)*; this is necessary for film to fulfil its potential as an art form. If everyone migrates to watching films on tiny personal screens, we will have lost something we need.)[14]

Film necessarily involves a gaze. The point is for that gaze *not* to be spectatorial, nor escapist, but to have instead the kind of character just indicated.

Film is (thus) a specifically relevant medium through which to think about moral issues. For, as philosophers such as Wittgenstein, Murdoch and Cora Diamond have taught us, how we look, what we see as what (and in what setting), is essential to who we will be, how we will act.

Some of the chapters that follow emphasize this by way of looking closely for instance at the representation and inhabitation and alteration of point of view in the films investigated.[15] Of course similar issues and possibilities are quite often present in literature; five of the films presented in the present work were in one way or another based on literary texts,[16] and I will often essay some comparison between the two forms. I make the case *en passant* that the old saying that "The book is always better than the film" is false. I think that that claim is true of many films, but I think we see some of the exceptions here. I think there is a good case to be made that some of these films that we are going to look at are on balance better than the books they are based on: in particular, *The Lord of the Rings* (trilogy), *Solaris* and *2001: A Space Odyssey*. Others are pretty clearly superior in certain important respects, even if not necessarily overall: here, I am thinking of *The Road* and *Never Let Me Go*, both of which are based upon very fine books indeed, but both of which, as I discuss, add something significant in their filmicness.

Books are like shared dreams, shared between author and reader. Films however are triply shared: for, on top of their being shared dreams as books are, they are shared in their *creation* (as the shared vision of director, editor, cameraman, actors etc.);[17] and their *viewers* share them together, *in a cinema*. (As already implied above, many good films are more or less explicitly interested in that very sharedness: including, in this book, as I shall describe, most obviously *Waltz With Bashir*, *Melancholia*, *2001: A Space Odyssey* and *Avatar*; and, elsewhere, such major works as Bergman's *The Magic Flute*, *Inception*, *Fight Club*,[18] *Prospero's Books* and *Rear Window*.)

Let me enter into a couple of ways in which film can achieve things that books can't. (And when those things are done well, a film can be better than the book it was based on.)

One thing that films can do that books can't is to give you the perspective of one of the characters, "literally," via a point-of-view shot.

Now of course books that are written in the first person can sometimes give you the perspective of that person through the narrator. But films can do that *too*, with the use of a narratorial voiceover (a technique used powerfully in some of the films present in this book). What books cannot do, obviously, is *show* you what the character is actually seeing. What I think we find in a number of the films in this book is that some moments where we see from the perspective of one or some of the characters or whatever are very important. (There are sometimes even point-of-view shots which are not from the point of view of *characters* in the film: I dwell on a fascinatingly brilliant one in Chapter 6, in which we encounter a point-of-view shot from the point of view of the One Ring.)[19]

Another thing that films can do that books can't (at least, not in the same way), is to break the "fourth wall." Doing so "alienates" the audience so that they no longer think they are watching something real. Such alienation is generally avoided, except in art films where it is exploited as a device, "following" Brecht. But if it's done in a certain kind of thoughtful way, the breaking of the fourth wall can actually *involve* you in the film in a way that books find it very difficult to do. Actors staring directly into the camera can seem somehow to say to one, "You cannot have us performing for you as a mere spectacle. We are (sometimes, such as *now*) reaching out to you, involving you. Like it or not." *This* mode of breaking of the fourth wall can be immensely creative, or compelling. It can "compel" the audience, *by* releasing viewers from the story-ness of the film, and re-inserting them into the actual world, the world that the film has screened, and ultimately disclosed. When this happens, then alienation happens *in order for return*. In order, at a deeper level, that is, to re-integrate, and *de*-alienate.[20]

Consider the following example (to which I'll return in Chapter 6). This is arguably a tremendous case of both of these phenomena that I've just described, together: an example, that is, of what I would suggest is perhaps a point-of-view shot which *becomes* very definitely a breaking of the fourth wall.[21]

In the final scene of *Avatar* the lead character, Jake, is lying down, comatose, as he seeks to transfer from his human body into his avatar body, and his lover, Neytiri, is tending to him in the crucial moment of the transfer. At the climactic moment she leans in over him and the camera follows and proceeds with what we can perhaps take to be her position as she moves to look into his eyes. Suddenly his eyes flick open—and gaze directly into ours.

This breaking of the fourth wall in the final moments of the film is directed at us, at you-the-viewer. My "reading" of this is that it constitutes a kind of call upon you to do the same thing as he is doing. Namely, to cease your complacent slumber. Metaphorically, *really waking up* is an awakening, a form of enlightenment. Waking up to reality. Which requires in turn waking up from the film into what the film invokes (or requires of us).[22]

I'd also argue, albeit less certainly, that that shot can be viewed as a point-of-view shot. The blue Na'vi woman, Neytiri, is leaning over her love. You the viewer are positioned, arguably, in *her* (awakened) point of view, the native point of view, the rebel point of view, the loving point of view:[23] as you then see Jake breaking the fourth wall; as you see him looking directly out of the screen. At you.

It is a striking fact that a number of the films in this book involve such breakings of the fourth wall:

- The ending of *Avatar* surely references the magnificent ending of *2001*, discussed in Chapter 5, in which the "Star Child" looks directly at the viewer, seemingly asking one, as I see it, to seek to enact (at the level of metaphor) the kind of rebirth that the film depicts.
- A striking moment in *Gravity* occurs when, floating inside one of the spaceships she passes through *en route* to completing *her* odyssey by returning home to Earth, the protagonist Ryan Stone (Sandra Bullock) takes a turn directly towards the camera. Directly towards *us*. (Before reversing to take the route she actually needs to take.) This moment is complemented later by a Brechtian effect in the very final moment of the film, when Stone hesitantly steps on the Earth's surface, and splashes

some mud onto the lens of the camera. (It is not coincidental that this kind of effect is often found at the very end of a film, at the time when it is often most useful to "prompt" the audience to make a connection between what they are seeing and the world we inhabit.)[24]

- *Melancholia opens* with precisely such a moment. Justine stares dolefully directly at one.[25] (Perhaps this—*opening* with such a breaking of the fourth wall—should be considered a more drastic version of the same kind of move as I just noted, parenthetically: a more drastic prompt to consider oneself *addressed* by the film, and reminded that one's absorption in it should not be equated with a permanent withdrawal from the world, but on the contrary.)

Then there are more moments where characters looking at each other get directly entered into, from both sides:

- As I shall discuss in Chapter 1, the very end of *Waltz With Bashir* demands to be read in something like this way, as the Palestinian women approach the soldier-protagonist, and we switch from one of their points of view directly to the other. These switches in visual point of view in the final two minutes of the film[26] are an invitation to enter into the point of view of the families of the massacred Palestinians *and* then into that of the Israeli soldier-protagonist who finally understands/remembers what he saw and was complicit in.
- Something similar is (arguably) true of the notorious "Not by force!" scene in *Last Year in Marienbad.*

The moments described here highlight or exemplify something important about the films under discussion. These films demand clear thinking, but they demand also something more meaningful still, perhaps something more unusual than that. They demand *consciousness*. Awareness. They demand—invoke, *create*—presence. The films certainly do not just provide moving images to gawp or marvel at. This is the ultimate reason why it would be a deep mistake to see these films as pure "spectatorship" or escapism. Rather, they matter. For they are about what matters not mainly in the sense of factively informing us about it but in the sense of enabling us to *see* (it), as if for the first time. And, as we see, and as we feel, so to change.

These "movies" are there to *move* us. And they and their ilk might even yet be levers with which to move our world to a better place.

What we are being asked to see? One might put it thus: we are being asked, precisely, to see. Not one thing in particular, but all. We are being asked to see holistically, ecologically. We are being asked to see what we constantly overlook, or refuse to notice. As Wittgenstein famously put it, thuswise:

> The aspects of things that are most important for us are hidden because of their simplicity and familiarity. (One is unable to notice

something—because it is always before one's eyes.) . . . [W]e fail to be struck by what, once seen, is most striking and most powerful.
(*Philosophical Investigations* [*PI*] 129)

What are we awakening from? We are awakening from misconceptions:

- About ourselves, our reality and our relations with others (and about those others *being* beings, being Being).
- About what is important.
- About what is right and wrong.

Perhaps sometimes, we are also awakening even from thinking itself. What can I mean by *that*? There is a bias that tends to exist in the Western world, especially the academic world: a bias in favour of thinking. We do philosophy because we are interested in thinking and are good at it. And (so?) we tend to think that it's the most important thing. This is a way that philosophers especially are prone to think, because what do philosophers do better than think? (In many cases, what else can philosophers do well, at all?) No wonder then, that philosophers tend often to denigrate the body and nature. This denigration has helped bring the world to the pretty pass in which we now find ourselves.[27]

Is there a way in which we are trapped, not just in a specific assumption or bias, but trapped within *thinking itself*? This is an extremely challenging idea,[28] because it goes against the grain of thinkers to consider that maybe thinking is not ultimately the aim, the need, the answer.

This book is not just about challenging or changing your mind (or even your perceptions about film[s]) but is about something bigger still. Most of the films on which I will dwell in the chapters to follow can potentially facilitate an awakening *from mind*.

An awakening into a more *present* state. A stirring and growing, *inter alia*, of your soul or heart (and then perhaps a changing of the world). The kind of awakening that I'm talking about, one that is presenced in every single film discussed in this book, is, I put it to you, badly needed now, at this dark time in our history.

Wittgenstein, as so often, gives an entrée into this. Startlingly for a philosopher, sometimes, at key moments, he specifically warns us not to *think*, but rather to do what moviegoers do: to *look*, to *see*. Here is an epochal example of the way in which Wittgenstein seeks precisely to free us from the tyranny of the "musts" that automatic thinking imposes upon us (He seeks here to enable us to contemplate that we might consider some important concepts as being what he will call "family-resemblance concepts," rather than "classical" concepts, [as] standardly [since Socrates] considered in philosophy):

Consider . . . the proceedings that we call "games". . . . What is common to them all?—Don't say: "There *must* be something common or

> they would not be called 'games'."—but *look and see* whether there is anything common to all. —For if you look at them you will not see something that is common to all, but similarities, relationships, and a whole series of them at that. To repeat, don't think, but look!.
>
> (*PI* 66)

I would like "film-theorists" and "film-philosophers" to *look* more at the films that they treat of or dissect. Rather than treating films essentially as mere exemplifications of pre-formed philosophical ideas, or as ideology waiting for the theorist to unmask it, or as mere raw material for projection onto of one's own ideas, I want to help us think *beyond* thinking, or certainly beyond all heresies of paraphrase.

I think that, when we actually *look* at (these) great films, we see the world aright.

Because the *cinematic awakenings* that follow yield necessary contemporary forms of timeless wisdom.

I now introduce the chapters that follow.

1. Implicating the Narrator, Implicating the Audience: *Waltz With Bashir* and *Apocalypto*

Waltz With Bashir has been considered both by several critics and by some parts of the public (especially, among activists) as an apologia for Israeli foreign policy and military actions. This chapter, while in sympathy with the toughest viable criticism of Israel, offers an opposite perspective on the film, pointing out how the same themes and techniques which have been taken to show the film's allegedly dismissive stance on Palestinians are in fact ways of first inhabiting and then turning beyond—in fact, challenging or even condemning—Israeli dismissiveness of Palestinians, as expressed in the murderous military operations featured in the film. The de-realizing, estranging atmosphere through which the protagonist as well as the viewer observes the unfolding of the 1982 Lebanon war serves, I argue, not *in the end* to distance from the events, but on the contrary to *show* how perpetrators of (war) crimes are able to do what they do precisely because they fail to fully acknowledge the reality of their victims. (That is, *they* distance themselves from the events that they are perpetrating.) Disturbingly and smartly, the film lulls one into a false sense of security through going along with such distanciation, for most of its duration. *Waltz With Bashir* thus constitutes a new version, ethically and politically charged, viscerally inhabited, of the concept of "unreliable narration."

A key to understanding the film along these lines can be found in the two scenes around which the film pivots. Firstly, the opening scene, which concerns a soldier who cannot peacefully sleep because of a recurring dream (which the film throws us straight into) about the dogs he killed during the war. Secondly, a scene halfway through the film in which the narrator's perception begins to shift when he finds himself overwhelmed by (a scene of) dead and dying horses, their suffering caused by an Israeli attack. These two scenes do not merely treat the animals they centrally concern as subhuman, or as metaphors: they illustrate the possibility of waking up to the reality of suffering outside the confines of "us" (humans). They are symmetrical with the shattering realization at the end of the film of the reality of the (being and) suffering of those human beings (the Arabs) who the film has, up to this point, been complicit in occluding. Thus the film wakes its audience up to the reality of Palestinian humanness by way of showing us their dehumanization. By *way* of alerting us to the danger of treating non-humans as if they are disposable. As Adorno put it, "Auschwitz begins when one says [of animals]: 'They're only animals.'"

The analysis undertaken here explains what is otherwise mysterious: *why* this film is an animation. The animation (and music) serve the purpose of *expressing* the disturbing distance Israeli soldiers have put between themselves and the Lebanese/Palestinian/Arab people. Until, at the end of the film, these victims, who previously were mere drawings, and whose voice one never heard, suddenly explode onto the screen and into one's consciousness with full power, reality and voice. As beings in their own right, who demand our acknowledgement.

The film's strategy displays a deep similarity with Mel Gibson's underrated *Apocalypto*. In the final minutes of *Apocalypto*, one comes to realize with a shock that what one took to be a critique of a cruel alien "barbarian" imperialism actually functions as a critique of a still crueller home-grown imperialism: the ecologically and culturally destructive European forces that invaded the Americas, i.e. "*Us*."[29]

2. How to Represent a Past One Would Rather Forget: *Hiroshima Mon Amour* and *Last Year in Marienbad*

Hiroshima Mon Amour demands to be read as a philosophical dialogue, from the very beginning. The film opens in an utterly unrealistic manner, with "Lui's" repeated denials of "Elle's" experience and memory in particular apparently making no sense; unless they are understood as having a philosophical (as opposed to an ordinary) point. This marks the "register" in which it is suggested here one should receive this remarkable work, a film that endeavours to enable one to understand how it might be possible to move on: both from a situation of having been (however indirectly) complicit in an unprecedentedly deep wrong (e.g. the dropping of the atom bomb, or the threatening to use such a bomb), and, similarly, from a situation of deep grief. The world of the film is one where, by means of an indirection, the possibility of near-total destruction of the planet, by means of technology, is *made real* both to the character and to the viewer, encouraging both to grapple with the ramifications of such an "unthinkable" possibility.

Despite appearances, the film does not really have characters. The chapter develops this point by focusing in detail on the ending of the film, where, in scenes that have not properly been understood, previously, it becomes evident that the film does not ultimately concern two individuals—the characters being in the end just a vehicle for enabling the viewer to experience the topos of the film more deeply—but rather, two cities, two nations, or two cultures. The real point of Resnais's film is to enable the past, however painful, to be remembered and reconciled with. The suggestion is that only if we do this will we not forget enough to be ready to engage in nuclear war again, which would bring an end to everything. The film seeks to represent, more accurately than a documentary might be able to,[30] the seemingly inconceivable but actual horror of what humanity did to itself and to the Earth, in World War II. It could help open our eyes to the danger of the ultimate ecocidal crime: a full-scale nuclear war (that would lead to a "nuclear winter").

When *Hiroshima Mon Amour* is experienced and understood in this way, the meaning of Resnais's other great masterpiece, the notorious "puzzle film" *Last Year in Marienbad*, starts to become clearer. *Marienbad* is presented here as "variations on *Hiroshima*'s theme": as a film which explores what happens when, rather than undergoing a process of "therapy" or reconciliation, one *remains* stuck inside one's mind, or in the past, forever. *Lost* in thought; stuck in a rational trap.

3. Learning From Conceptually Impossible Versions of Our World: *Never Let Me Go* and *The Road*

This chapter shows how *Never Let Me Go* is to be understood as what Wittgenstein calls *an object of comparison*: its differences from the actual world, as well as its similarities, are profoundly instructive. The film is an exercise in pushing a boat out further than it can actually be pushed. In watching this heart-destroying film, one leaves the bounds of sense. The central claim of the chapter is that the "society" shown in the film is conceptually impossible. The main aspect of this impossibility is that the human "clones" have no thought of rebelling: the film shows a world in which humans are *constitutively* unfree. By comparing such a society to our own, as the film encourages us to do, we can gain some profoundly important insights about our constitutive freedom (and about how lightly we often abandon it). As in Wittgenstein's "philosophy of nonsense":[31] one looks back in from this impossible "outside," and learns something about sense and about oneself/our social and (increasingly un-)natural world.[32]

Never Let Me Go pushes against the limits of possibility in a manner so extreme as to be deeply revealing. A similar ability to present instructive impossibilities is found in *The Road*, which this chapter presents as minor key to *Never Let Me Go*, because *The Road* offers a conceptually impossible post-apocalyptic world in which the biosphere is dead but we are not. It invites us to explore what it would be like to inhabit that world, and be defeated in such exploration, in order that a new relationship to our world is made possible.

Both films challenge the viewer's imaginative capabilities in the service of increasing our intellectual autonomy; which is a process of liberation. In this way, *The Road* and *Never Let Me Go enact* film as philosophical "therapy"[33] as liberation, in anyone willing to listen, *willing* to watch, to see. The problem here is (once more following Wittgenstein) a problem of the *will*, not primarily of the intellect. A philosophical capacity, and thereby a political one, that is lacking in all of the protagonists in *Never Let Me Go*. By contrast, the utterly grim world of *The Road* nevertheless facilitates a surprisingly redemptive ending, which focuses on the protagonist's surprising encounter, at last, with something non-human (albeit domesticated) and alive: a (potentially friendly) dog.

4. When Melancholia Is Exactly What Is Called For: *Melancholia* and *Solaris*

This chapter explores von Trier's *Melancholia* as a richly intertextual work, referencing and involving previous great films on psychopathology, including Resnais's *Last Year in Marienbad* and Tarkovsky's *Solaris*. The latter similarly considers a (profoundly) alien planet as a metaphor for our own unhappiness and self-destructiveness. The ways in which *Melancholia* explicitly or implicitly references *Solaris* at key points are investigated, including how, while Kelvin (the protagonist of *Solaris*) lives while Justine (the melancholic of the title) dies, he lives in illusion, while she, I submit, dies in presence. *Melancholia* makes its own novel intervention by thinking "ecopsychologically": that is, by intervening in the vital debates around human-caused climate change (potential slow cataclysm through sea-level rise, overheat etc.), our potential destruction of our living home (the Earth), *and* how our psychology, while feeding into this, might also yet offer a route out of it. Justine, the film's first and foremost protagonist, personifies this intervention: she is maladapted to the everyday world but well adapted to a world facing destruction. The planet Melancholia in the film is a mega-object come to shatter our certainties and complacencies.[34]

The greatest contribution of the film is made in the closing ten minutes. These vividly show how, with the complete bracketing of normal expectations of life going on, when facing resolutely what Heidegger calls one's (our) "being towards death," out of the jaws of destruction we can reclaim a determination to connect and to live, to experience each moment of life, with others. *Melancholia* offers a shattering experience of shattering and reconnection. It is a work that reconnects[35] us (to one another, to life itself, to what makes our lives possible) at a time in planetary history when nothing could be more important. By reflecting on the film's portrayal of depression and destruction and "resurrection," the chapter argues that *Melancholia* is in fact a stunningly acute "reminder" of the value of life and thus brings out our life-affirming ability, in extreme situations, to connect with the world and other people. The possible total destruction of our posterity by our present short-sightedness is desperately sad. Surely we *need* to feel both shame and melancholia about what we are doing to our world and ourselves? We need to mourn the world we are in the process of destroying—*if* we are to stop destroying it.

5. Gravity's Arc: Or *Gravity: A Space Odyssey*

This chapter ties closely together the hero's journeys of *2001: A Space Odyssey* and *Gravity*, arguing that they share a common kinship with a more or less Arendtian dissatisfaction with the technophilic desire to leave Earth behind; and thus that they midwife an ecological reawakening.

Close to the heart of this chapter is the lengthy, superb, necessary tracking shot that begins in the 13th minute of Cuarón's *Gravity*.[36] The shot is an aesthetic move and a highly educational one, because it involves a (repeated, slow, impossible) entry into the "point of view" of the film's main protagonist, as she undergoes the crisis of an accident in space (caused by the reckless pollution of space by spacecraft-wreckage) around which the film's plot is based. This "long take" simultaneously provides an invitation to reflect upon the meaning of that movement in and out of her point of view.

The transformative experience of *Gravity*, the chapter suggests, lies in its invitation to move out from Earth only to return to Earth—after, that is, having explored and abandoned the attraction of an existence, made visible in the film, detached from natural "earth(l)y" conditions. The film restores us as—essentially—earthlings. The last portion of the chapter explores the sense in which *Gravity* is clearly *inter alia* a reworking of *2001: A Space Odyssey*: above all, in being necessarily a voyage and return, that necessarily involves a rebirth. What goes up must come down. Who goes out must come back. *Gravity* focuses (us) upon the vital sense in which life itself is only possible here, on our home, the beautiful and beloved Earth.

6. The Fantasy of Absolute Safety Through Absolute Power: *The Lord of the Rings* Trilogy and *Avatar*

This is by a long way the longest chapter in the book. That's first because it deals with a trilogy, but second and more important because, in dealing with the highest grossing films of all time, it meets the ultimate test of a central claim of mine: that hugely popular works can be philosophically significant, and may even be popular partly *because* of that significance, that meaningful dialogical journey.

It may seem self-evident that *The Lord of the Rings* is a (would-be) *pre*-Modern text. But the interpretation of Tolkien's work manifest in Peter Jackson's film version suggests another possibility: that *The Lord of the Rings* may be viewed rather as a psycho-political meditation on madness,[37] allied in certain respects to Modernistic and to some extent even to post-Modernistic understandings of such madness (principally Louis Sass's as well as, in certain respects, Wittgenstein's, Derrida's and Foucault's). The desire to put on the One Ring could helpfully be seen as a metaphor for the alienated (and self-defeating) effort, in our eco-destructive and individualist capitalist culture, to achieve an invulnerable individual "inner" sense of safety, power and control. (In this regard, the interpretation of *Lord of the Rings* offered is broadly inspired by Buddhism.)

What the Ring actually produces or induces, however, is despair—or rather, it presents one with the full *attractions* of despair. What is worked through in Jackson's films, the chapter claims, is the non-actuality of despair as part of a human life. So long as one is thinking/being/acting at all, one is *not* in despair: Frodo never despairs of Gollum, nor even quite of himself. This, the chapter claims, is the surprisingly contemporary teaching of *The Lord of the Rings*, deeply consonant with the (roughly contemporaneous) work of Kafka and Primo Levi. Tolkien's fictional camps and marches and would-be exterminations are no more a cause for a lived *despair* than are even Levi's utterly harrowing non-fictional counterparts thereof. Even at their lowest, human beings such as Levi and the wearer of the Ring always remained human. Thus the three films that make up the trilogy lead viewers on a journey that constitutes the overcoming of the delusive temptations of a powerful phantasm: despair, masquerading as a lived/liveable phenomenon. The chapter explores this as the films do, in a highly accessible way, by following the hero's journey, which is ultimately a psychical (and cyclical) one. Moreover, I explore here the (highly salient) context of ecology-under-threat in which the journey takes place, especially in *The Two Towers* (with reference to Isengard).

A lengthy "coda" to the chapter compares *Lord of the Rings* to the neo-mythic epic that followed, the greatest box office success in cinematic history: *Avatar*. *Avatar* punctures the quest for safety-through-anti-ecological-economic-and-military-power. Just as in the *Lord of the Rings* trilogy, it does so by making a Jamesian/Kierkegaardian/Pascalian move: a leap of faith. The conclusions to both films involve a military defeat that nevertheless enables a different kind of victory (in *Avatar*, one brought about through the impossible uprising of Pandora's non-human animals). The conclusion of *Avatar* in particular invites careful thought about how we can seek to win the struggle for ecology and survival (because Gaia is not going to ride to *our* rescue). One reason why the *Lord of the Rings* trilogy and *Avatar* have been so popular is that they take audiences to and out of dark places of a kind that we are actually creating on Earth, today, and in this sense they are the *opposite* of what they are too often assumed to be: escapism.

Conclusion: What Have We Learnt?

Philosophical liberation necessarily involves a change in perspective on the world. The political, contemporary relevance of the films discussed in this book is the central object of reflection here. They in effect bring out the pressing nature of recent (political and) ecological developments and disasters. The real question may be: Can films help wake us up *in time*? What have we learnt or could we learn (from these films); have we learnt enough; and can the learning be shared quickly *and* deeply enough?

The Conclusion, however, also brings out something else: that, while I have tended in this book to emphasize the importance of the macrocosm—of attitudinal change at the societal level, of *political* change, of the growth and enaction of ecological consciousness—this is necessarily the flip side of microcosmic change. And that has been strikingly clear in some of the films focal in this book, in various ways (re-explored briefly here). So the book ends by noting how in the end liberation and ecology are as much about mental health as they are about democratic change. Enlightenment in its true sense is a healing of the self inextricable from a healing of the world.[38]

Notes

1. Green Screen (Exeter: University of Exeter Press, 2005) by David Ingram began this trend; that work of cultural studies does a good job in placing and problematizing most of the films it discusses. More recent works of ecocriticism have gone further still, helpfully questioning anthropocentrism while analyzing films: exemplary titles here are Ecocinema Theory and Practice (edited by Stephen Rust, Salma Monani and Sean Cubitt; London: Routledge, 2013) and Screening Nature (edited by Anat Pick and Guinevere Narraway; Oxford: Berghahn, 2013). Paula Willoquet-Maricondi's Framing the World (2010) shares some central goals of this book, i.e. making viewers of film more ecologically minded and responsive to the environment. Ecology and Popular Film (by Robin Murray and Joseph Heuman; Albany NY: SUNY, 2009) also explores Nature in film, although only as represented mainstream films.

   However, "eco" film studies books, such as the ones mentioned here, typically still take a spectatorial stance towards the films they discuss, often an ideological stance; because they frequently treat the films that they are discussing *as*, allegedly, samples of ideology. The approach of the present work is strikingly different. This book takes it that the films discussed have something to *teach* us and a possibility of *transforming* us, which is brought out by analyzing the experience of viewing the films though the lens of broadly ethical and existential concepts. This book, then, is above all film *as* philosophy; or, perhaps better, film-ecosophy.
2. As it is for Wittgenstein; on which, see my Liberatory Philosophy (forthcoming).
3. In Philosophical Investigations (London: MacMillan, 1958 (1953); henceforth PI) 123.

   The way that we can gather a Wittgensteinian orientation to these films can be compared to what Wittgenstein says at PI 527. Here, he seeks to understanding what it is to understand a theme in an artwork. One says, in relation to the work, things like "Don't you see, this is as if a conclusion were being drawn", or "This is as it were a parenthesis". In this book, I repeatedly use strategies of Wittgenstein's philosophising in just this way: i.e. I suggest that *they are profoundly analogous to strategies of the films* that I am engaging with. This starts in Chapter 1, with my suggestion that the method of *Waltz with Bashir* is akin to the method of the Tractatus.
4. This structural device is inspired by the original edition of Stephen Mulhall's splendid book, On Film (2002). Mulhall was my teacher in Wittgenstein, and

the origination of my interest in film-as-philosophy came when I read, at age 20, the draft version of his groundbreaking essay on *Blade Runner*, "Picturing the human, body and soul." (Cf. also below.)

5. Robert Hunziker, "Hidden danger of ecological collapse", www.adbusters.org/news/hidden-danger-ecological-collapse/; Damian Carrington, "Earth's sixth mass extinction event underway, scientists warn", www.theguardian.com/environment/2017/jul/10/earths-sixth-mass-extinction-event-already-underway-scientists-warn; Damian Carrington, "Earth has lost half its wildlife in the last 40 years", www.theguardian.com/environment/2014/sep/29/earth-lost-50-wildlife-in-40-years-wwf.
6. See William Ophuls, Immoderate Greatness: Why Civilisations Fail (London: CreateSpace, 2012), and E. O. Wilson, Half-Earth: Our Planet's Fight for Life (New York: Liveright, 2016). See also my talk here www.youtube.com/watch?v=xkshpb4-gAU.
7. See my "The forward march of the Greens halted?", www.newstatesman.com/politics/elections/2015/09/if-corbyn-becomes-leader-whats-left-greens, on what lies beyond the "left vs. right" opposition. See John Blewitt and Ray Cunningham (eds.), The Post-Growth Project (London: LPP, 2014), for the outdatedness of—still hegemonic—growthism.
8. On the normalcy bias in academia *vis-à-vis* climate, see Kevin Anderson's "Duality in climate science", https://kevinanderson.info/blog/duality-in-climate-science/. On how academia (and most of the rest of society) tends to be unable to think holistically or to comprehend the enormity of the hole we are in, see Iain McGilchrist's The Master and His Emissary (New Haven: Yale, 2009).
9. On how desperate our situation is, a very good clue is provided by the attitudes of those who are slightly awake among the super-rich: see e.g. Douglas Rushkoff's "Survival of the richest", https://medium.com/s/futurehuman/survival-of-the-richest-9ef6cddd0cc1 and Mark O'Connell's "Why Silicon Valley billionaires are preparing for the apocalypse in New Zealand", www.theguardian.com/news/2018/feb/15/why-silicon-valley-billionaires-are-prepping-for-the-apocalypse-in-new-zealand.

   On the desperateness specifically of our climate situation, see e.g. David Wallace-Wells's "The uninhabitable Earth", http://nymag.com/daily/intelligencer/2017/07/climate-change-earth-too-hot-for-humans-annotated.html and climate scientist Prof. Kevin Anderson's talk here www.youtube.com/watch?v=zjTtohMgGk8&t=1212s.

   On why we are not going to "solve" the climate crisis, see Green House's "Facing up to climate reality", www.greenhousethinktank.org/uploads/4/8/3/2/48324387/intro_to_fucr_project_2017_spring_conference_edition.pdf; and my "Climate change: once we no longer deny it then we might just have the will to change course", www.thelondoneconomic.com/opinion/climate-change-once-we-no-longer-deny-it-then-we-just-might-have-the-will-to-try-drastically-to-change-course/14/03/.

   On why there is still hope, see my "A case for genuine hope in the fact of climate disaster", www.thelondoneconomic.com/opinion/a-case-for-genuine-hope-in-the-face-of-climate-disaster/09/03/—and read the book you are reading . . .
10. For explication of why, see Bryan G. Norton's "Convergence and divergence: The convergence hypothesis twenty years later," in Ben A. Minteer (ed.), Nature in Common?: Environmental Ethics and the Contested Foundations of Environmental Policy (Philadelphia: Temple University Press, 2009).
11. And in this connection I owe an enormous debt to my former editorial assistant, my former student Dr. Silvia Panizza, who is one of the world experts on Murdoch's philosophy.
12. See Stephen Mulhall's reading thereof, a kind of founding document of contemporary film as philosophy: http://xroads.virginia.edu/~DRBR2/mulhall.html, especially the first couple of pages.

13. See my "The tale Parfit tells: Analytic metaphysics of personal identity vs. Wittgensteinian film and literature," *Philosophy and Literature* 39:1 (Apr. 2015), 128–153, for a take on how *The Prestige* is quite explicitly a film about our attention in a cinema.
14. Thanks to Julian Hanich for enabling me to understand fully the importance of this point.
15. The general background to my emphasis on point-of-view shots can be found in George Wilson's Narration in Light: Studies in Cinematic Point of View (Baltimore: John Hopkins University Press, 1986).
16. Additionally, the two Resnais films I focus on were both "authored" (or co-authored) by high-art authors not known as screenwriters/scriptwriters.
17. Cf. www.theawl.com/2010/07/the-key-to-inception-its-a-movie-about-making-movies, on *Inception* as a shared dream from its inception: i.e. as a film about film-making.
18. Nancy Bauer's essay in my Film as Philosophy (Basingstoke: Palgrave, 2005) collection reads *Fight Club* in this way.
19. Films also sometimes intriguingly give you the point of view literally of cameras/of viewers: as in the uncanny opening scenes of *The Truman Show*. (Compare also the way that Antonioni does something similar, in periodically—especially in *The Passenger* and *L'avventura*—making one fairly starkly aware of camera movement and position and orientation.)
20. I argue, in the closing pages of my "Wittgenstein as unreliable narrator/unreliable author," in Ana Falcato and Antonio Cardiello (eds.), Philosophy in the Condition of Modernism (Berlin: Springer, 2018), that this is part of what is occurring in Bergman's *Persona* and in *Fight Club*, at both the "object" and "meta" levels. I.e. that these two films stage a re-integration from a state of psychological dis-integration, and that they do so in part *by way of* highlighting their own filmic nature, i.e. by way of "alienation," especially at the end (and start) of these films. I claim that *Fight Club* is a kind of remake—or, rather, a (good) *interpretation*—of *Persona*.
21. This moment is of course not from a (film from a) book, but makes nevertheless the point that I'm making about what films can do that books can't do: www.youtube.com/watch?v=-S0Lc6hZ9mk.
22. Cf. the notes above: thus there is a sense in which any film, to achieve something extra-textual, is best-advised at some point to be clear about its own artificiality.
23. As we'll see in Chapter 6, this is precisely the key change in point of view that the film as a whole seeks to execute.
24. Think for instance once more of the very end of Bergman's *Persona*.
25. This might helpfully be compared with other beautiful such fourth-wall breakings (and, more generally, alienation effects) in Trier's films: such as Bess's magnificent smilings at the camera in *Breaking the Waves* (or the famous catching of one camera by another in *The Idiots*).
26. Go 1 hour 21 minutes into it: www.youtube.com/watch?v=ynH68E1GEdc.
27. See my "Some thoughts on civilizational succession", www.truthandpower.com/rupert-read-some-thoughts-on-civilisational-succession/ and my "Climate change: once we no longer deny it then we might just have the will to change course", www.thelondoneconomic.com/opinion/climate-change-once-we-no-longer-deny-it-then-we-just-might-have-the-will-to-try-drastically-to-change-course/14/03/, for some (further) salient detail on the extreme desperateness of our current situation, about which virtually all of us are virtually always in at least soft denial.
28. An idea I shall find, specifically, in *Last Year in Marienbad*.
29. Assuming that the main audience of this book, as of *Apocalypto*, is "Westerners," not indigenous etc. people. After all, as I point out in discussing *Avatar* in Chapter 6, it is not *they* who need to learn and change, but us. The early part

of *Apocalypto*, and the Na'Vi in *Avatar*, offer *models* of (social and) ecological embeddedness.

30. Here of course there is a direct connect with *Waltz With Bashir*: in both cases, the ordinary nature of documentary is sublated, to produce something that achieves the aim of documentary, emotionally and *epistemologically*, better, probably, than any "straight" documentary could.
31. For explication, see my and Crary's edited book, The New Wittgenstein (London: Routledge, 2000), *passim*.
32. This central difficulty that one is forced to wrestle with is represented both by the scientific advances which have led to the creation of human clones, which force the question of what "genuine" humanity consists of, and, more important still, by the way the clones are used as "mere means" for medical purposes.
33. Cf. Garry Hagberg's "On philosophy as therapy," *Philosophy and Literature*, 27:1 (2003), 196–210, especially p. 203.
34. It is thus clearly, among other things, a metaphor for dangerous human-triggered climate change. In this way, it is comparable to eco-philosopher Timothy Morton's conception of climate change as a "hyper-object."
35. Cf. Joanna Macy & Molly Brown, Coming back to life: The updated guide to The Work that Reconnects (Gabriola Island: New Society, 2014).
36. Whenever I give timings of frames/scenes in this book, I am using the timings of British DVDs: i.e. timings based on the film running at 25 frames per second.
37. Of various kinds, but especially of schizophrenia spectrum conditions: cf. e.g. my "On approaching schizophrenia via Wittgenstein," *Philosophical Psychology*, 14:4 (2001), 499–514.
38. Thanks to Peter Kramer and Naomi Marghaleet for very helpful editorial comments on earlier versions of this Introduction.

1 Implicating the Narrator, Implicating the Audience

Waltz With Bashir (and *Apocalypto*)

> Auschwitz begins wherever someone looks at a slaughterhouse and thinks: they're only animals.
>
> —Theodor Adorno

Waltz With Bashir: A *political* film, clearly.[1] But *philosophical*? How so? How am I saying that this film belongs in this book?

A first approximation would be: because it is a film about acknowledgement and/*of* reality. But to be able to answer the question properly, let me begin apparently less directly, by dwelling on some of the political criticism that this film has received.

The film is said by "Leftist" critics to be an apologia for the military adventurism of Israeli "liberal humanism," a paean to the capacity of Israel to covertly pat itself on the back by means of self-criticism. It is said to be merely a therapeutic exercise for the continuing perpetrators of violence. It is said to involve a failure to humanize the Arab victims of the first Lebanese war. The film is also said by some, notably by Shohini Chaudhuri,[2] to be *de*politicizing: an indulgent, *merely* therapeutic fleeing from history and from political reality.

These are serious criticisms. By my means of refuting them,[3] I will seek to reveal the sense in which the film succeeds in being a philosophical work.

The mode in which I will refute these criticisms might be unexpected. I don't seek to refute them "head-on." Indeed, I concede outright that there is a sense in which the reading of the film that these criticisms encapsulate is to some considerable extent a natural reading of the film. (What the critics are talking about is *very much* "in" the film; in something like the same way that the picture-picture of meaning is "in" Wittgenstein's *Tractatus Logico-Philosophicus*.[4] The question is whether this presence is simply the promulgation of a picture or view, or whether it turns out to be an indirect way of carrying out a profound *criticism* of that view. In the case of the *Tractatus*, most of what is "in" the book, *within* the "*frame*" of the book, is profoundly criticized, and overcome, by its end.)[5]

Clearly, some Israelis themselves saw the film in this way; that is, as almost a sort of celebration of Israeli military history. The very things that the critics homed in on appear to be the very things in the film that many of the film's more jingoistic fans/supporters welcomed. I accept that it is possible that the film has by and large had this kind of effect. Just as it is possible that the effect of Pink Floyd's *The Wall* was to license an attitude among a generation of schoolkids to the effect that education was (is) evil. But that reading of the Alan Parker / Pink Floyd film was at best a gross oversimplification, at worst a complete failure to understand the film (which was rather a critique more specifically of *schooling*, in something like Ivan Illich's sense of that word,[6] and more generally a critique of hierarchical mass-conformism). *A fortiori*, while it may be true that *Waltz With Bashir* has been widely "understood," by foes and fans alike, to be an apologia for Israel,[7] that does *not* prove that that is the right way to take the film. And I shall explain why it is not.

What exactly do I mean by saying that it is to a fair extent nevertheless a "natural" reading of the film? Well, the film is made, starkly, from the Israeli point of view. One never hears an Arab speak in the whole animation; the Arabs in it are mostly just bystanders, enemies or victims/cannon-fodder. The film is absolutely explicitly a search for therapy and memory and personal peace on the part of Israeli ex-soldiers, especially the protagonist. I "concede" all of this. Much more than that: I claim that this is precisely *how* the film does its work.

In this regard, *Waltz* might be helpfully compared to *In the Company of Men*: a film which proceeds strikingly from the male (masculinist, macho) point of view, which it inhabits explicitly, relentlessly and unpleasantly. It would be a profound failure of vision to understand that film as a celebration of male domination, when it is the very opposite. The claim that I shall make about *Waltz With Bashir* is similar.

(Incidentally, it follows from this that the "Bechdel test" for determining whether a film is sexist is at best an oversimplification with limited scope of application. *In the Company of Men* would fail the Bechdel test, just as *Waltz With Bashir* would fail a similar test *vis-à-vis* Orientalism / taking Arab voices seriously. But what I have just shown is that any such test is blind to the possibility that a film is a worthwhile—and perhaps devastating—critique of the oppressors' worldview, *from the inside. Most* of the films in this book in fact fail the Bechdel test [the only exceptions are *Melancholia* and *Never Let Me Go*]. I think that that proves that the test itself fails.)

Chaudhuri observes correctly that "The way [*Waltz With Bashir*] maps space is through the aggressor's perspective, through the grids of power, surveillance and control."[8] Well yes, indeed: How is one going to dismantle the self-exculpating mindset of aggressors more effectively than by means of inhabiting it from the inside and bringing to full self-consciousness the horror of it, when it is brought face to face with its results?

Chaudhuri continues: "Carmi [one of the protagonist-filmmaker's fellow soldiers] . . . relates how he and his comrades used to fire indiscriminately, at whom they knew not. In the visualized recollection, they continue shooting relentlessly when a Mercedes comes into view, riddling it with bullets until the door opens and a dead Arab flops out." Yes indeed; the film shows starkly Israeli responsibility for criminal killings of civilians, and evokes the dehumanization that made this possible. Weirdly, Chaudhuri seems to think this (scene) some kind of would-be exculpatory celebration of unreasoned almost genocidal violence, when it is quite plain that the scene in question is putting in your face the nature of this dehumanization: all animated, all "deniable," but all (too) real.

"Suffering is not really suffering when it is drawn in lines," asserts Gideon Levy.[9] Yes, indeed: it is easier to take it as somehow fictive, or as somehow not deep or human suffering. *That's the point.* The film, I will claim, is an animation *because* it aims to convey the semi-wilful suspension of belief/acknowledgement that enabled Israel to do its dehumanizing "work" in Lebanon. It aims to manifest to us the lived "acid trip" experience of a war whose wagers were trying to make it somehow deniable, uncriminal. The film's critics think that the film is an animation so as to be able with finality to deny the humanity of the oppressed in it. A more complete misunderstanding of the film would be hard to imagine.

So, indeed, the film, through virtually its entirely length, precisely and deliberately fails to humanize the quasi-generic victims who are most appallingly and casually dispatched. It does this, in songs like "I Bombed Beirut Today," with a knowing nod to *Apocalypse Now*; it foregrounds the willingness of the Israelis to delight in killing and to be carefree about "collateral damage."[10]

Chaudhuri notes the link to *Apocalypse Now*, but, incredibly, fails to see this as evidence of *Waltz With Bashir*'s clear intention to denounce rather than to celebrate the apocalypse wreaked by aggressors (then, the American invaders; now, the Israeli invaders). Instead, she remarks that the sequence in question "inadvertently conveys the sheer arrogance of the war: Lebanon is reduced to a playground where Israeli soldiers can indulge their libidinal fascinations with their war machines."[11] As so often in her essay, Chaudhuri here gets things just about exactly right, EXCEPT for the crucial fact that she claims, bizarrely, that the effect is "inadvertent." The film is shot through with war crimes, with innocent Arabs needlessly and wilfully shot through, and it centrally concerns the protagonist's emerging and finally clear guilt at all this: How can this possibly be "inadvertent"??

The film follows intently the protagonist's and his former colleagues' potentially self-indulgent but profoundly understandable—in most cases, their minds, their souls are not at peace, because of what they have done, because of what they are guilty of—search for personal peace in the wake of all this. What it finally enables us to understand is how this search will be fruitless unless one actually wakes up to and starts to actually come to terms with one's status as a guilty party, as a perpetrator, as a collaborator with the kind of horror in fact that was inflicted on the Jews in the Holocaust.

What the film shows us, searingly, is how the Israelis that it depicts saw—or rather, for far far too long *failed* to see—those that they oppressed, senselessly murdered, committed to torture and death at the hands of their allies etc. The film shows us the routine nature of Israeli war crimes, the total complicity with the greatest war crime of them all in that war (Sabra and Shatila)—and the unwillingness to see/acknowledge these things. It takes the need for therapy on the part of traumatized Israeli victims and shows how, at least in the case of Folman himself, this need was itself a product of that unwillingness. Folman can't remember what happened to him during the war, and especially at Sabra and Shatila—because he can't bear to see what he did. Who he was. What he was a part of. Only a long and increasingly painful (but also potentially/ultimately liberating)[12] journey enables him to see (this).

The film explicitly and precisely concerns the tendency to dissociate, to treat what one is doing to others as merely something that one is watching. Again and again the film shows us soldiers inclining to view the war as if it were a spectator sport, or *a film*.[13] Just as one can watch a film, and think that one is simply "escaping." Such a false *fantasy* of freedom is precisely what the book of mine that you are reading is designed to oppose. *Waltz With Bashir* subjects one to (a) more uncomfortable *reality*.

Consider in this connection Gideon Levy's influential criticism of the film:

> This is an extraordinarily infuriating film precisely because it is done with so much talent. Art has been recruited here for an operation of deceit. The war has been painted with soft, caressing colours—as in comic books. Even the blood is amazingly aesthetic, and suffering is not really suffering when it is drawn in lines.[14]

But the *point* of the film, on the understanding of it that I'm developing here, is exactly to lull the audience into those soft caressing colours, that aesthetic. To get them (us) to be (too) comfortable in a world drawn in lines. And then gradually to realize, especially at the close ('at the death,' as one might helpfully put it), with a horror of recognition and criticism at oneself too for having been complicit with it,[15] how one has gone along with a dehumanization, or "Orientalization," of the Arab victims.[16]

This is the beauty of the shift in register in the closing scenes of the film (that I'll discuss in detail below). Point-of-view shots are used in the film periodically, but none more powerfully than those with which it closes. We see as if we were the Palestinian women streaming out of the camp, Folman standing before us; and then we flip around to see from his point of view, almost literally: as the camera held by an Israeli documents the reality of the women's extreme grief at the extraordinary crime that has just been perpetrated against them.

Thus the biggest criticism of all made of the film, that it allegedly involves a failure to see Arabs, that it can't even take them seriously enough to depict them as other than cartoon characters, is in the end the very *inverse* of the truth. The film is not an *instance* of that failure; it is precisely an *uncovering* of all such failure. Up virtually to the very end, one could potentially make that criticism, at a great stretch, but at the very end, with the irruption of reality onto the screen,[17] the rug is pulled from under one: suddenly, one comes in full colour to *see* the failure to see the other that has been endemic throughout the film. This is flashback with a vengeance: this is suddenly true documentary as opposed to quasi-fictionalized/"animated" documentary.

This is *why* the film was made as an animation: *because* of the way that it is designed to midwife a "therapeutic," liberatory journey through the danger of pseudo-acknowledgement to the moral truth and real acknowledgement. The film, as an animation, evokes in its very warp and weft the dissociation from reality that the film repeatedly thematizes as a feature of the Israeli invasion of Lebanon and as a basis for the trauma that the surviving invaders went on to suffer from. The soldiers at the time were detached from the reality of what they inflicted, saw and endured; and this effect is only heightened in retrospect, in the suppression of traumatic memories and guilt.

This surreality or derealization, this would-be escape from reality, is what the film aims to excavate and to manifest the overcoming of, in a vivid "return of the repressed." The dominant would-be criticisms of the film are not, it turns out, criticisms of the film: they are criticisms made *by* the film. The film's "critics" have taken the bait that the film offers, swallowed it whole as if it were simply food like any other food, and then complained that all they can see or feel is a bare hook. When what they should have done is realize just how difficult it is to see bait *as bait*. Just how difficult (but nevertheless possible) it is to practice "therapy" in a non-corrupt way, a non-self-indulgent, non-egotistical way. (That is my project in this book.) The film seeks to show us how extremely deep (as deep as the unconscious, as deep as the sea which is so often in shot or in the background, in the Lebanon scenes of the film) the resistance of the protagonist and his colleagues was to seeing what they had done, to acknowledging reality. And how this is the case for most of us, most of the time, if we are honest about it.[18]

The film is an animation in order to capture and manifest the degree to which the Israelis saw the whole thing as a kind of video game,[19] the way in which they distanced themselves from the cruel reality they were bringing about. The protagonists' dissociated take on the war that is shown us in the film is not an excuse for murder:[20] the film is exactly showing us how deep a *problem* it is, a disastrous failure to see. A disaster for which everyone who allowed themselves to take part in or to facilitate such murder bears responsibility: though the film correctly tends to place the greatest responsibility on the Israeli leadership (as well as on the Phalangists who actually exterminated the Palestinians in the Sabra and Shatila camps).[21]

The film precisely, deliberately obscures the humanity of the Palestinians and the Lebanese—the Arabs—from us, until the veil is ripped away, at the end. This is no denial of responsibility on the part of the soldiers who are the film's characters: on the contrary (though those critics who, simplistically and unwisely, are so determinedly angry that they want to place Folman on exactly the same level as Sharon/Begin and the Phalangists will be disappointed).

To one not watching with enough attention, the film seems to be alleging throughout its course that it is simply documenting reality. This makes its political critics angry, as they claim that this documentation is one-sided, involving a sheer failure to see the side of the oppressed and massacred. But again, that is the very point that *the film itself* is making! Whereas a film such as *The Lives of Others* does this by enabling us to follow in detail the change-in-view of an oppressor who painfully comes to realize that that is what he is, until he manages to stop being it any longer, *Waltz With Bashir* accomplishes this task of "rehumanization"—of the aggressed-against, and thus ultimately of the aggressor[22]—by a more painful means: the actual cinematic inhabitation of the oppressor's mindset, until the guilt can actually be genuinely acknowledged as deserved, the memories that have been repressed retrieved, the illusion of the lesser humanity of the victims exploded.

Specifically, this film makes sense of the repression of memory suffered by its protagonist as a denial of something so painful and so important that it took a process of years of deliberate work to get it back: what Folman was in denial about was that those who he and his whole army was complicit in the massacre of were ordinary human beings. People. This truism that yet somehow one can forget under the burden of a depoliticizing propagandistic ideology, of nationalism, of deliberate desensitization, is what comes crashing in on him at the end. Suddenly, the faces of the wailing women we have seen repeatedly during the film, drawn crudely, are alive.[23] Suddenly, their reality is unavoidable, their grief unbearable. At least, to those who have not—as it seems so many of those who rushed to welcome the film as a vindication of Israel *or* who rushed to condemn it as an apologia for Israel *have*—in effect stopped watching by (or long before) that point. Refused to really look and see. Thus the supreme irony is that the very phenomenon that the film is criticizing is revealed in the reactions of many of the film's "critics" *and* "supporters."

The guilt of the characters has itself been used against the film. A particularly mean-spirited moment in Chaudhuri's critique is her remark that "[I]n *Waltz,* we are summoned to ratify Folman's guilt (*or rather, lack of it*) regarding the Sabra and Shatila massacres."[24] This remark is telling: Chaudhuri cannot really make up her mind whether or not the film features guilt, nor whether or not that is a good thing. She periodically makes remarks that appear to suggest that even feeling guilty is somehow a wrong (because somehow self-indulgent) response for a perpetrator to have (but, what else are Folman et al. supposed to feel?). The film is *about* the guilt that Folman and some (not all—some of the film's dramatis personae don't exactly come out of it smelling of roses) of his colleagues feel: a striking example being Boaz's dream, with which the film begins, of the dogs that he has killed coming back to haunt him (to kill him). Chaudhuri twists this into Boaz allegedly being cast by the film as the victim (of these would-be murderous dogs), but it is entirely clear that the dream comes from Boaz's guilt about having so fatally treated these innocent beings. And it would hardly be a stretch from there to thinking of resonant *clichés* (which have of course a vast and great literary provenance, from Kafka and Malcolm Lowry to Coetzee) such as being treated or indeed shot "like a dog," and to thinking quite literally of the dehumanization of Arabs (in Lebanon, and by extension in the Nakba) and of Jews (in the Holocaust) that the film takes such a striking interest in. Boaz's dream stands proxy for guilt at how the *Arabs* in the film / at the time were treated. A mode of treatment that Israelis above all ought to know better than to mete out to anyone.[25]

The widespread resistance to the film—resistance to what it is, I am suggesting, really *about*, and therefore also the very resistance that it *is* about—suggests a deeply unfortunate unwillingness to enter into the subtlety of "unreliable narration" as manifest in *Waltz With Bashir*. The film's narration is unreliable, in the sense that we never *hear* the truth, not in the whole film. We are offered snatches of it, more and more fragments of it, but never enough. We are offered the opportunity to *see* it, finally, at the very end, when at last Folman, the protagonist, is able to see what lay in plain view but what he has repressed all these years: that the people at Sabra and Shatila were indeed people, not "terrorists," not enemies.[26] But ordinary human beings worthy of respect and care. That the women were fellow human beings, in desperate grief; that the bodies lying there were bodies of human beings; that the dead little girl buried almost up to her nose in rubble was a little girl just like any other little girl, such as an Israeli little girl. I say we are *offered* this opportunity: again, very ironically, it appears that the critics of *Waltz With Bashir* are not themselves prone to take up the offer. They cannot comprehend the extraordinary way in which the humanity of the victims has been disclosed by the film, via its deliberately slow-cure, byzantine, indirect, *therapeutic* route.

The film draws a bold, explicit parallel,[27] almost never admitted into polite Israeli society normally, between the Israelis' fear of genocide—what almost "successfully" was done to the Jewish people under the Nazis (who, in trying to cover up their actions, were of course the original "Holocaust-deniers")—and the suppression of the reality of the Sabra and Shatila massacre: the unwillingness to admit, to acknowledge, that that was a quasi-genocidal action that the Israeli military and political leadership in particular directly facilitated the perpetration of. A parallel is drawn quite explicitly between the camps in World War II and the camps "now" (the refugee camps where Palestinians were slaughtered in Lebanon). This is hardly denialism; this is hardly being soft on Israel. On the contrary. Chaudhuri seeks to counter this point by interpreting Folman's therapist, Ori, as arguing that Folman need not feel guilty for what he did at Sabra and Shatila, because the guilt is somehow a product of "inherited" Holocaust-trauma.[28] This seems to me a questionable reading of the scene. For, even if it were a correct reading of Ori's words (which is far from self-evident), it would not help Chaudhuri's case: because Folman looks quizzically at Ori, as so often he looks quizzically at interlocutors in the film when they in one way or another seek to minimize responsibility for what happened. Folman does not seem to accept what Ori says. Ori's stance (if it is as Chaudhuri claims it to be) is not logical: the film has clearly signalled to us that there *is* a remarkable degree of moral equivalency between what guards at a concentration or extermination camp did and what Folman et al. did at Beirut . . . because, in particular, the film makes clear that Israeli soldiers, including Folman, *did* help facilitate the massacre (in particular, by illuminating the camps by flares, so that the Phalangists could see what they were doing). His recovery of that memory is an absolutely key moment in Folman's coming to terms with his own truly guilty past.

Let me add a final respect in which one can see, if one allows oneself to actually see *the film*, that it is about the process of acknowledgement, and its difficulty. That it is about the difficulty of achieving a therapy that actually changes something and doesn't merely return one to the front line to kill again. I am referring to the film's repeated interest not only in human victims, but in animals, as well as (more broadly, in the background of the film as it were, as it witnesses the gradual wreaking of more and more damage on the invaded country) in the destruction of landscape, heritage and ecology.

There are two moments in the film that strongly feature non-human animals. They are crucial to the film.

As noted above, the film *begins* with a blatant and dark "return of the repressed." It begins, that is, with the would-be revenge, in dream, of the many dogs killed by one of the Israeli soldiers featured in the film. He killed these dogs because he was seeking to shut them up before his fellow soldiers moved in to occupy a village. He feels guilt at taking the lives of these innocent creatures.

Later, of course, we come to see that many innocent Palestinians, too, were slaughtered "like animals." These dogs can be seen as metaphors for their human "owners."

But they can also be seen as more than metaphors. They can be seen, too, as *themselves*. One might, that is, come to see the killing of the dogs as itself a real crime. Relative to which the killing of innocent people is doubtless a still *greater* crime. I am suggesting that the film seeks ultimately to give one eyes to see both.

The thought that that is how one ought to see, according to this film, is supported by a moment that is I think the turning point of the film.[29] Folman is struggling to understand his dissociation, his inability to place himself in the events that he experienced during the Lebanon invasion. A therapist tells him of a similar dissociation that she knew of, in a soldier who was also in the 1982–3 assault on Lebanon. This soldier got through the war by imagining that he was watching his experiences *on film*: "*He looked at everything as if through an imaginary camera*," or "*as if watching the war on film*." Then one day, suddenly, "his camera broke." He (suddenly) couldn't any longer fail to *see* what was in front of his eyes. Its awful reality. What was the event that smashed his "camera"? It was seeing the aftermath of an Israeli attack, in which many Arabian *horses* were left dead and dying, horribly wounded.

A critic might say: "Ha, you see?!! He couldn't see Arabs as real. But he saw horses, finally, as real. This soldier in effect thinks that Arabs are less than animals!" And this critic might well think that he has found "yet another" proof of the alleged tacit racism of the film.

There might be something to the criticism. It is sometimes easier to see animals as fellow beings than it is to see demonized "othered" humans as fellow beings.

But the criticism would go horribly wrong if it were meant as some kind of dismissal. For the "camera" "breaking" is to be welcomed: and once it is broken, then one cannot help seeing. All. Just as Folman at the very end of the film suddenly cannot avoid seeing the wailing women careering towards him as real, as humans, as victimized fellow beings. As he said earlier in the film, "Maybe I'll discover things I don't really want to." This is indeed what has happened, by the end.

It is not good enough to sneer at the seeing of horses as fellow beings. For they *are*. And they get seen as what they are far more rarely, in the round; "our" animals are subject to a constant horrific use and destruction.

What is so wonderful about this heart-rending horses scene in the film, that I have called its turning point—for it is from this point onward that the film's protagonist starts to see through the mechanism of denial that he has been implicated in, and becomes fated to have to remember and to *see* the other, and no longer to be in a state of dissociation—is of course the metaphor that the soldier used, that the therapist relays: that of seeing as if *on film*. What *Waltz With Bashir* is a film that doubles its filmicness by being an animation, and in a sense triples it by repeatedly showing its characters trying to avoid seeing what they were seeing, by seeing it as *merely* screened. (And thus, they screened reality from themselves, by their own spectatorship.) They were trying, that is, rather to treat their experience as if it were "merely" a film. What this film, *Waltz With Bashir*, shows us is how film can be the very opposite of *merely* film. How it can actually disclose reality, by cutting through the-desire-to-distantiate-reality-by-treating-reality-as-if-it-were-something-one-was-merely-spectating-upon, as one might well pre-analytically imagine the watching of films to be. A spectator sport. Whereas in this book, I am arguing that film (like philosophy, and like human life at large) is at its best about the very reverse—about awakening to reality, which always means that one is part of this world, not just gazing at it as if from outside. This realization comes to us, finally, drastically, in the transition from animation to documentary footage, at the very end of *Waltz With Bashir*.

This film in fact discloses reality most focally, by disclosing the reality of our unwillingness to have reality disclosed to us.

So, what if I am right? What late lessons can we then learn about—or from—*Waltz With Bashir*?

Well, we can make sense at last of its *title*. The title appears to pick out one striking but odd—extreme and specific—incident in the film, where an Israeli soldier "dances" around a street, firing wildly at hidden Arab snipers, miraculously surviving, and appearing as he does so to be dancing with the murdered Lebanese Christian leader whose portrait was ubiquitous, and whose killing was apparently the spark that lit the genocidal revenge of the Phalangists in Sabra and Shatila. Why does this deserve to become the title of the whole film?

My suggestion is that the reason is reasonably clear, once one understands the analysis I have essayed above. What the film, once one understands its end, *means* is that Israelis haven't wanted to face reality: the reality of the horrific experiences that many of their soldiers had in the war; specifically, the reality of the war crimes they were engaged in in Lebanon (which haunt them); the reality of their complicity with the Phalangists, their appalling permission/facilitation of the massacre at Sabra and Shatila;[30] and most crucial of all, underlying all of these: the reality of the Palestinians and of the Lebanese. This film is utterly about *acknowledgement*, in the Cavellian (and Mulhallian) sense of that word.[31] It is focally about the failure to acknowledge the humanity of "the other": the failure of the Israelis to *see* the Arabs. In other words, the Israelis have been unwilling to acknowledge their own waltz with Bashir: (culminating in) their own severe degree of complicity with his dedicated militia that carried out the Sabra and Shatila massacres. They have denied their own dance with—their dalliance with, their indulgence of—murderous exterminationism and war criminality. The film is *about* that reality, and that failure of acknowledgement.

Wittgenstein spoke of one's task in philosophy as one akin to therapy, to psychotherapeutic methods (*Philosophical Investigations* [*PI*] 133). He drew on Augustine's *Confessions* to begin that task (*PI* 1). He wanted to help one to accomplish for oneself a confession, a realization, a therapy. He thought that this was how our civilization might change for the better and become something worth calling cultured: if enough of us were willing to undertake this self-therapy, as an ethical task of coming to honesty and integrity, of working to overcome delusions that one didn't really want to overcome. I am of course not claiming that *Waltz With Bashir* is as great a (philosophical) work as Wittgenstein's *Philosophical Investigations*! And Folman's film, playing as it did with such difficult, terrible material, probably ought to have gone even further towards accepting the truth of what Israel facilitated at Sabra and Shatila, and of what Israel did to Lebanon. But I do think that there is a genuine family resemblance between the film and Wittgenstein's philosophical intent. Just as Wittgenstein has been catastrophically misunderstood, so has Folman's film. Chaudhuri sneers at the film by calling it confessional and therapeutic,[32] as if those were terms of criticism. Taken correctly, they are exactly not. They are routes towards beginning an ethical transformation, one that issues (in Folman's case) in a true and important confession, a genuine acknowledgement of complicity, of guilt in its true sense. The film culminates in a screened "declaration," for those whose eyes are open, of the constant previous failure to *see* the victims, to draw them as anything other than cartoon characters. Indeed, this culmination even contains a "coded" call for the filmmaker to do these true victims' bidding. The words in Arabic that the wailing grieving Arab woman, finally seen in real life, is saying, as she walks towards the camera, translate as "Film, film and send the images to foreign countries!" *And this is what Folman has done.* He has found a way that might actually be effective[33]—harder to ignore, harder not to inhabit, because of the way it gets inside one, because of the discomfort forced upon one by inhabiting the aggressors' shoes, and acknowledging one's closeness to them as our "allies"—of dissolving the Orientalist illusion of separateness between us (the film's presumed audience: the film is intended to wake up "Westerners" including Israelis, rather than Arabs, who are already in most cases all too aware) and the real victims. He has filmed what is so hard to see, the oppressors' mindset, *and* the dissolution and destruction of that mindset, and he has "sent" the film, including the very footage that that Palestinian woman wanted shared, to the countries that need to see this.

Waltz With Bashir is a film that can free one from the prejudices that are manifest in racism and oppression, violent subjugation and not-so-obviously violent othering. But one has to overcome these prejudices actively; one has to be willing to open one's eyes. One has, so to speak, to *throw away* the ladder of the animation, once one has climbed up it. Only then, can one—does one, *will* one—acknowledge reality.

***

To conclude this essay, I will essay a comparison with another foreign-language film about war and empire and our animal kin and our ecology that came out around the same time, and that, while not earning as much criticism as *Waltz*, has been perhaps similarly under-rated and under-understood. I am referring to *Apocalypto*.

Why make the comparison? For a very specific reason: *Apocalypto* too is a film that, I argue, *functions* by sucking the viewer in, "encouraging" them to take up a perspective that, at the end of the film, is suddenly pulled from under one's feet.

The whole *weight* of *Waltz*, I have argued, becomes apparent only in the final minute or two of the film. Those who have rushed to judgement, condemning the film, have closed their eyes, in effect, by then. Perhaps my essay might re-open them a little.

The weightiest part of *Apocalypto* takes up more screen time. But not much more: it is in the final four minutes of the film. For the entirety of the film up until this point, one's sympathies have been with the "native" village that gets so cruelly destroyed and enslaved by the "Mayans," the savage imperial overlord "civilization" in the body of the film.[34] In the latter half of the film, one is caught up in the unlikely rush to freedom (and to saving his wife and child[ren]) of the film's protagonist, Jaguar Paw.[35] Finally, in the last four minutes of the film, Jaguar Paw stumbles, to be saved by an utterly unexpected, incomprehensible (to him and his pursuers) event: the arrival on the coast there in the Americas of ships bearing Europeans. Colonizers. Enslavers who will turn the savage civilization of the Mayans into a nothing, and will wreak a new level of ecological and human destruction.

Suddenly one realizes what the film has been about. It is not really about the cruelty of a conveniently distant "barbarian" Mayan imperium and the destruction of the cultures it encountered. It is about us. *Waltz With Bashir* ultimately facilitates the true horror of realizing that one has been complicit in dehumanizing the victims; *Apocalypto* goes even a step further, and flips completely who the oppressed and oppressors are (going to be).

For, more specifically, *Apocalypto* is about the unwillingness of empires and their members to see themselves as they (we) really are. It is *about* our failure, in the course of watching the film up to that point, to realize that the film was a perfect metaphor for ourselves. For *our* destruction alike of nature and of other cultures. (A destruction pre-figured in their small way by the Mayans of the film. At the hands of their [by our standards, petty] industry, we are told, and see, "The Earth bleeds.")[36]

Apocalypto delivers a devastating critique of the oppressors' worldview, *from the inside*—because we didn't realize (until the end) that we were inside it.

Just as *Waltz* ultimately thematizes one's failure to critique deeply enough the Israeli imperium in Lebanon/Palestine, and allows one at last to see the victims as humans, so *Apocalypto* suddenly turns around one's hatred of the Mayan oppressors and allows one to see them as victims-to-be, and thus as humans, too. The most important part of this process is the way that it enforces upon one (or at least facilitates) a thorough self-criticism. One's self-satisfaction as one watched the film, one's easy identifications and dis-identifications, are suddenly thrown over. And one learns something about one's desire or willingness to fail to think deeply enough, to fail to challenge assumptions that help one to feel good about oneself: the very desire that Wittgenstein identified as his number one target of criticism in philosophy.[37]

Thus *Apocalypto* delivers to us its grave warning.

Wittgenstein once remarked, in *Culture and Value* (C&V) that:

> The truly apocalyptic view of the world is that things do *not* repeat themselves. It isn't absurd, e.g., to believe that the age of science and technology is the beginning of the end for humanity; that the idea of great progress is delusion, along with the idea that the truth will ultimately be known; that there is nothing good or desirable about scientific knowledge and that mankind, in seeking it, is falling into a trap. It is by no means obvious that this is not how things are.[38]

Indeed, it is quite possible that this is how things are.[39] Assuming that they are not so is hubristic, and thus potentially reckless.

What is the meaning of Wittgenstein's first sentence here? Against the view that history runs in circles/cycles comes the possibility that, if *our* civilization collapses, it will bring down the entire future with it. For our collapse will most likely be a result of a terminal breaching of the *global* limits to growth (perhaps through atomic warfare, perhaps through runaway climate change that leads over a century or so to climate-apocalypse). The truly apocalyptic view then is that we don't/won't get the chance in future even to repeat (what by then will be) past mistakes.[40]

What films like *Apocalypto* and other films discussed in this book give us is the chance to learn from past mistakes before it is too late. If we can dare not only to seek to recapture some of the wisdom of the indigenous,[41] who have in many cases inherited their predecessors' learnings from previous ecological disasters,[42] but also (and this is the really hard part) to recognize ourselves in the Mayans of this film, then it might not be too late. Such films aim to give us a chance to be *able* to make future mistakes, a necessary condition for which is to not eliminate ourselves and our beautiful home altogether.

Waltz With Bashir asks us to learn from the appalling history of Sabra and Shatila, so that we don't repeat history. In particular, it asks us to learn the most difficult thing about this history: namely, how enduringly difficult it is, as Westerners, to detach ourselves sufficiently from the perpetrators in order to really be able to acknowledge adequately the victims. How difficult it was for the filmmaker to do so, and how we travelled with him, or else (if we condemned the film and refused to really see it) forestalled that travelling by condemning up front in a way that doesn't acknowledge the reality: that *ordinary men*[43] not unlike us facilitate and carry out such incredible crimes. A too-easy condemnation of *Waltz* lets oneself too quickly off the hook.

Apocalypto asks us to learn from the (often appalling, ecologically as well as humanly) history of the pre-European American Meso- and South American empires *and* from the (even worse) history of their destruction (and of ongoing degradation) by our ancestors and their descendants, the colonizers of the Americas . . . so that we don't repeat it. In particular, it asks us to learn the most difficult thing about this history: namely, how enduringly difficult it is to notice that it seems to be repeating, and to take responsibility for that. *Apocalypto* is about—or rather, it gives us an *experience* of—how our civilization will destroy itself from within: *unless* we learn from films such as this/these.

Apocalypto gives us a kind of cartoon version of Meso-American empire; we find ourselves very easily and willingly appalled by these "barbaric" baddies. *Waltz* literally gives us a cartoon version of the Israeli invasion of Lebanon; in particular, it cartoonizes the Palestinians and the Lebanese. And its "Left" critics wish to *attach* to a cartoonish version of the Israelis. But both films overturn all these cartoons, in their dying moments.

The failure of many viewers/readers of *Waltz With Bashir* to acknowledge the "lesson" that I have suggested the film actually yields, their intent simply to condemn, is itself highly suggestive: often such failures are overdetermined by unacknowledged desires. Could there be a moralistic political element to such failure, such resistance, such as the desire to just code Israelis as all inevitably "bad guys" (and to see oneself as nothing whatsoever to do with them and nothing whatsoever like them), not wanting ever to risk empathizing with them, in order to keep one's political-moral categories neat and comfortable?[44] We would be wiser to acknowledge the reality that acknowledging the reality and humanity of others—victims *and* killers—is difficult, and can never be taken for granted. It is an achievement devoutly to be wished, and needing deliberate effort. And indeed courage.

To sum up: *Waltz With Bashir* has been considered by both several critics and some ('right-on') parts of the public as an apologia for Israeli foreign policy and military actions. This chapter, while in sympathy with the very toughest (viable) criticism of Israel, has offered an opposite perspective, pointing out how the same techniques which have been taken to show the film's allegedly dismissive stance on Palestinians in fact turn out to be subtler, more effective, more genuinely "therapeutic" ways of condemning Israeli dismissiveness of Palestinians, including as expressed in the murderous military operations featured in the film. The de-realizing, estranging atmosphere through which the protagonist as well as the viewer observe the unfolding of the 1982 Lebanon war serves, this chapter has argued, not *in the end* to distance one from the events, but on the contrary to show how perpetrators of crimes are able to do what they do precisely because they fail to fully acknowledge the reality of their victims. Disturbingly and smartly, the film lulls one into a false sense of security through going along with such distancing, for most of its duration. In this way, it could be seen helpfully as embodying a new, ethically and politically charged, version of the concept of "unreliable narration." In the end, the narrator gets implicated—and the unaware viewer with him.

This filmic strategy displays a similarity, I closed by suggesting, with Mel Gibson's *Apocalypto*. In the final minutes of *Apocalypto*, the audience comes alive to its own implication in the whole motion of the film. One comes to realize with a shock, that is, that what one took to be a critique of a cruel alien imperialism actually functions as a critique of a still crueller home-grown imperialism: the ecologically and culturally destructive European forces that invaded the Americas and took over the whole globe, whose legates we are. One's sense of safety and of complacent identification with the victims is swept away, and one is left with something much more challenging and unsettling, forcing one to think again about one's own place in the world—and about our responsibilities to preserving this beautiful place of ours.[45]

Such rethinking—and refeeling—via revealing and revisioning, is what this book is all about.

Notes

1. Though let me say right off the bat that I still find it disappointing to see that people want the film to be a political *essay*. It should correspond, point by point, they think, to what "we" (the politically right-on) think about the conflict. No film can do this. Or rather: no film even *should* do this, unless it sets out to be in effect an ideological-political pamphlet, rather than (as is really rather plain in the case of *Waltz With Bashir*) a personally inflected work of art, actively involving viewers in a philosophical, historical and political reflection. I take it that the failure to understand this film, and many others, stems from this mistake. (I owe the guts of this point to Odai Al-Zoubi [and Gary Francione], to whom, many thanks.)

   My own political views on Israeli foreign policy and the Palestinian question are similar to those of many of the film's critics: I am sceptical as to the viability of the two-state solution, given how far the Israeli occupation of Palestine has gone; I think serious consideration should be given to the one-state solution, which would dismantle Israel and its Bantustans and substitute a democratic bi-national state; to pressurize Israel to the negotiating table, I support sanctions on Israel, including economic and academic sanctions. But it would seem to me sheer barbarism—an absence of culture—for one to claim that a film ought only to reflect back to one exactly one's political views or prejudices. Thus, though doubtless Folman and I wouldn't see entirely eye to eye on politics (see n.3), I am interested in opening my eyes to see what may be good in his film. I would hope that others would be, too.
2. In her Cinema of the Dark Side (Edinburgh: Edinburgh University Press, 2014).
3. Although I allow that it is entirely possible that many viewers, both "foes" *and* "fans," have seen the film without seeing what I aim to bring out here. Those who think the film somehow justifies Israel or shows what a great state it is have *completely* missed the point: but it should be acknowledged that it is risky to make a film about such a difficult and political subject, and especially to make it in the way that Folman chose to (deliberately occluding the victims etc.). One opens oneself up to misunderstanding. But the point once more is that only by making a film which is nothing more than a straight documentary or a rather dogmatic political essay can one avoid this risk—see n.1, above. If only such films were to be (allowed to be?) made, then most of what is in this book would not even be possible. It would be an end to art.
4. London: Routledge, 1922.
5. For explication of this reading of the book, see Cora Diamond's The Realistic Spirit (Cambridge, MA: MIT, 1991) and my co-edited collections The New

Wittgenstein (London: Routledge, 2000) and Beyond the *Tractatus* Wars (London: Routledge, 2012). (See in particular Diamond's (and Conant's) notion of this kind of philosophy as essentially "indirect communication." Films-as-philosophy might be described using the same term. Contrast this with the way that a fully paraphrasable work would actually be a kind of direct communication. For one could just as well substitute the paraphrase, the message.)

6. See his classic Deschooling Society (New York: Harper & Row, 1971).
7. Some say that Folman's own remarks about the film nevertheless prove this. I don't accept this, mainly because my interest is in the film that was actually made, not in its "author's" subsequent remarks about it. But also because, for every problematic remark of Folman's about the film (e.g. his worrying seeming-claim in pre-release publicity that the film suggests that Israeli soldiers did not share responsibility for the Sabra and Shatila massacre: "Lest we forget", Jonathan Freedland, www.theguardian.com/film/2008/oct/25/waltz-with-bashir-ari-folman), not only—crucially—are such remarks contradicted *by the film itself* (which crucially makes clear, for instance, how the sending of flares up over the camps helped the Phalangists to carry out their massacre: Chaudhuri admits this crucial point at p. 153, but somehow this does not dent her confidence that the film is designed to enable Israel to slough off its guilt, rather than to come to terms with it); they are also contradicted I believe by *other* remarks made *by Folman himself* (e.g. in the interview with him which accompanies the DVD of the film; no one hearing this interview could be in any doubt about who Folman thinks were the perpetrators hereabouts, and who the victims), thus cancelling out the alleged "proof."
8. Chaudhuri, Cinema of the Dark Side, p. 152.
9. In a passage quoted with approval by Chaudhuri, Cinema of the Dark Side, p. 152.
10. See this scene, for instance: www.youtube.com/watch?v=EmBvRfZKDwM. See also www.youtube.com/watch?v=Jxv1Zs0gMMk.
11. Chaudhuri, Cinema of the Dark Side, p. 151.
12. I use the term "liberation" here in the ordinary sense, which is also the sense in which Wittgenstein uses it (on which, see my book Liberatory Philosophy, forthcoming, and the short discussion of this theme in the Introduction to the present work) when he says that what philosophy is centrally about is becoming free (of prejudice; of propaganda; of the continued attachment to one of the eggshells of thinking which one hopes to have grown out of). This is also close to the central Buddhist sense of "liberation."

    One's suspicion is that critics like Chaudhuri and Levy do not want the guilty to experience any liberation. They want the perpetrators to remain cowed and suffering in a perpetual state of would-be-atonement. This is an ugly desire; it is the desire for us all to remain stuck in history as in a nightmare, in an unbreakable pattern of anger and guilt. (Cf. Chapter 2.)

    We *need* perpetrators to face reality, obviously, just as, in Chapter 6, I'll explain why we need films that figure Westerners as becoming indigenous. Rather than as the dangerous, hegemonic "development" narrative would have it, the other way around.

    The oppressed don't need to change nearly as much as the oppressors do. (However, I'll also seek to explain, in Chapter 6, why it is often so psychologically attractive to remain oppressed, to keep oneself stuck in the "pure" and unagentic place of victimhood.)
13. See e.g. www.youtube.com/watch?v=WbV5Vf2TfEU.
14. "*Waltz with Bashir* is nothing but charade", at www.redpepper.org.uk/Anti-war-film-Waltz-with-Bashir-is/.

15. See www.youtube.com/watch?v=lgmjH1XbqXM.
16. Similar points apply to the harsh criticisms made of the film by Naira Antoun in her Review at http://electronicintifada.net/content/film-review-waltz-bashir/3547—I discuss the central such criticism in n.19, below.
17. See www.youtube.com/watch?v=lgmjH1XbqXM.
18. On this point, see Chapter 8 of my A Wittgensteinian Way With Paradoxes (New York: Lexington, 2013).
19. Chaudhuri writes (Cinema of the Dark Side, p. 157) that the film has a "logic of abstractification that dehumanizes Palestinians and holds their lives to little account," decrying the film's "animated war game aesthetic." We are now in a position to see how completely wrong-headed an accusation this is. Or rather: it is absolutely correct, as an account of what occurs IN the body of the film, often. Where it fails completely is in seeing that the film is "framed" by aesthetic choices—most strikingly of all, the last 90 seconds of the film—that show that what Chaudhuri criticizes is precisely what the film itself is criticizing. Chaudhuri's criticism is Folman's very own criticism—of what we often see *in* the body of the film (and similarly of what the world famously saw in the Wikileaks disclosures, in relation to the American dehumanization of Iraqis: https://www.youtube.com/watch?v=25EWUUBjPMo).

    Similar points apply to the critical review of *Waltz* published in *Electronic Intifada* (Naira Antoun, 19 Feb. 2009; http://electronicintifada.net/v2/article10322.shtml). Once again I find it astonishing how totally the point of the film is missed. "Palestinians are absent from the film," Antoun claims, in a voice of anger: yes, that's the POINT of the way the film is made! And of why it was made in the way it was. That's how the end of the film is shattering, as the protagonist wakes up to the fact that his guilt and nightmares are BECAUSE he had failed to see the Palestinians as people, and had seen them instead only as cartoon characters (or figures from in a shoot-'em-up video game). Because he'd tacitly absented them from humanity.
20. This appears to be how Chaudhuri reads it! (Cinema of the Dark Side, p. 154)
21. This is the explanation for the troubling moment in the pre-film publicity referred to in n.3, above: Folman wants to force Israel into reflection on its own responsibility for the war criminality it undertook and unleashed in Lebanon and Beirut, but it would be going too far to blame soldiers like Folman exactly as much for Sabra and Shatila as Sharon and the Phalangists should be blamed, and as much as the ordinary soldiers *should* be blamed for the murderous acts we see them at times undertaking themselves earlier in the film.
22. See my discussion of *The Lord of the Rings* in Chapter 6.
23. In this way, the end of *Waltz* is like a ramped-up version of the end of *The Corner*, the drama-documentary series that preceded *The Wire*. At the end of the final episode of *The Corner*, we suddenly meet the real live people who three of its main characters were very closely based upon: they materialize unexpectedly on our screen. The effect is dramatic, and moving. We realize what we knew all along, but somehow probably didn't quite fully allow ourselves to comprehend: that these awful, depressing stories and scenarios were true; the people portrayed in them real; their struggles, in many cases, ongoing.
24. Chaudhuri, Cinema of the Dark Side, p. 153, second emphasis added.
25. One should think here too of the work of Isaac Bashevis Singer, Jewish winner of the Nobel Prize for Literature, who wrote extensively about the Holocaust, and who held that animals were subject to an "eternal Treblinka."
26. Cf. n.61 and n.62 of Chapter 6, and *supra*.
27. Chaudhuri admits this (2014, p. 154). But she continues to insist at length that the film ought to be harder on the actual soldiers: Folman et al. I would have

thought that the more politically intelligent stance to have taken on this point would have been to have made clear the war criminality of many Israeli actions in its Lebanon war—which the film does—while putting the blame for this (as, again, the film does) more on the shoulders of the political and military leadership which issued these orders and unleashed the dogs of war than on the grunts who carried the orders out, following the script. Isn't it striking enough that the film explicitly blames Begin and Sharon for their criminal behaviour? Oughtn't they indeed to bear a heavier weight than the likes of Folman for what occurred in Lebanon under their watch?

28. Chaudhuri, Cinema of the Dark Side, p. 155.
29. It can be viewed here: www.youtube.com/watch?v=zFpCGI1abQA.
30. Robert Fisk, Pity the Nation (Oxford: OUP, 2001), helpfully explains how these points are connected. Fisk recounts in that book how he sought to tell Israeli soldiers who were at the gates of the camps while the massacre was peaking what was happening inside—and how they were unable or unwilling to believe that those inside the camps were other than "terrorists." Of course, what Folman / the viewer finally *sees*, at the end of the film, is how very far from the case this was. Finally, this reality of the victims crashes through the propagandistic ideology that has suffused the Israelis.
31. Beginning, that is, with Stanley Cavell's "Knowing and acknowledging," in his Must We Mean What We Say? (Cambridge: CUP, 2002 (1969), and continuing with Mulhall's "Picturing the human, body and soul."
32. Chaudhuri, Cinema of the Dark Side, p. 153.
33. Compare here the argument that I'll make in Chapter 2, with regard to Hiroshima (*mon amour*).
34. It is of course supremely irrelevant that the Mayans never actually, historically, encountered and were suborned and destroyed by European conquistadors. For the Aztecs and Incas, whose similarities to the Mayans in terms of what the film is interested in are quite sufficient, of course *were*. (In any case, the film doesn't explicitly identify its imperium as Mayan. It's just that that happens to be the closest fit of the Meso- and South American pre-European empires to that that we are shown in the film.)
35. The animal after whom he is named, and the other animals of the forest, destroy most of his Mayan pursuers, allowing him and the tiny remnants of his indigenous people the chance for a new start, symbolized in the film by a (literal) new birth. Thus, in his final words in the film: "We should go to the forest. And seek a new beginning." These words are not without relevance to us, too. (Cf. Chapter 6 on *Avatar*.)

    Apocalypto, unlike *Waltz* (and also unlike *Avatar*) is shot from the perspective of the oppressed. It is important that there are movies that are. But in the end what the three films have in common is far greater than what divides them. All three are ultimately urging upon us that perspective, and the terrifying realization (writ large in *Waltz*, as in *Apocalypto*) of what it actually amounts to stay complacently in the mindset of the oppressors *while fantasizing that one is being just to the oppressed*.
36. N.B.: Mayan civilization appears to have collapsed because of the unsustainable ecological pressure that it put on its environment: see Julie Kunen's "We and the Mayans share the same environmental concerns", www.theguardian.com/environment/blog/2012/dec/20/ancient-mayans-environmental-concerns-apocalypse.
37. See e.g. the first lecture of his Lectures on the Foundations of Mathematics (edited by Cora Diamond; Chicago: U. Chicago Press, 1975), and Lecture XI, wherein (on p. 103) Wittgenstein remarks on how his method goes "against the grain" of most of his listeners.

38. Wittgenstein (1980, p. 56e).
39. See e.g. the argument of Ronald Wright's A Short History of Progress (New York: Canongate, 2006). Cf. also John Gray's work, or Paul Kingsnorth's.
40. My thinking here is indebted to Tim Flannery's The Future Eaters (Kew: Reed Books, 1994). See also my own "Wittgenstein and the illusion of 'progress': On real politics and real philosophy in a world of technocracy", *Royal Institute of Philosophy Supplements* 78 (2016), 265–284.
41. For instance, as depicted especially in the early part of *Apocalypto*: their great sense of *temporality*, of ever-succeeding generations.
42. Here once more Flannery's argument in The Future Eaters is vital. Flannery resolves the paradox that many indigenous peoples live in a good deal of harmony with their environment, while their ancestors devastated their ecosystems and exterminated the megafauna. The resolution is: the survivors *learnt* from their disastrous mistakes. But we would not be able to learn from our mistakes if we were to collapse the entire global ecosystem.
43. I am thinking here of Browning's (London: HarperCollins, 2017 (1992)) extraordinary and telling book of that name, that concerns the everyday reality of the Holocaust, and the troubling way in which most of its perpetrators were not sadists or monsters, but just more or less ordinary men, who were helped to do the appalling things they did by mundane distancing devices such as getting drunk.
44. I owe this point to Silvia Panizza.
45. There is a deep similarity here with *Little Big Man*, the very ending of which (as we witness the title character now psychologically broken by what he has seen and recounted, the savage dismantling of American "Indian" societies) is devastating, in forcing one to take seriously what until that point has seemed sometimes a cartoonish and comedic history of violence. (The Dustin Hoffman protagonist in *Little Big Man* bears direct comparison too with Jake Sully, in *Avatar*.)

2 How to Represent a Past One Would Rather Forget

Hiroshima Mon Amour (and *Last Year in Marienbad*)

Deep grief is aversive. Grief repels entry, seeks even to repel understanding.[1]

When we suffer deeply, there is a desire to forget what we have experienced. This desire is perhaps redoubled, if we didn't *really* experience it; for instance, if other people suffered it "for" us, or if we denied it. If it is/seems just too awful to face. If we'd rather not have to acknowledge its reality.

But repression or denial of the past can, notoriously, lead to its (mistakes and horrors) returning, or being repeated.

We have to come to terms with what happened; and this includes with the scale of the loss.

***

How are we to understand the peculiar way in which *Hiroshima Mon Amour* begins?

We are presented with two lovers, intertwined, covered with what is perhaps radioactive dust. Then in the next shot the lovers are intertwined the same way, but thankfully without the dust. Then they speak to one another. But this is no ordinary lovers' discourse.

They speak about Hiroshima. "Elle," we later discover, is a French actress-visitor playing a part in a film being made about the catastrophe of Hiroshima and the response to it in Japan among peace-loving people. She explains how she has come to know about what happened in Hiroshima through a sequence of means: through learning various facts and figures; through visiting the site; through visiting survivors in hospital; through watching dramatized film reconstructions at the museum; and so forth. Her Japanese lover, "Lui," replies in a baffling way. To all her claims to "know," he replies in the negative. He disclaims/opposes her ordinary account. He says that she has seen nothing in Hiroshima.

What can he possibly mean?

He appears to be denying the validity of ordinary criteria. He seems to *deliberately* step outside their confines.

Now, "Wittgensteinians" or (in particular) advocates of "Ordinary Language Philosophy" sometimes say that the kind of thing he is saying is therefore without sense. Meaningless.

But that's too quick a judgement. Traversing the ordinary limits of sense is just what Lui means to do! So a claim that he is violating ordinary logical or philosophical grammar and is therefore speaking nonsense and can therefore simply be dismissed is simply no good. Lui surely *knows* he is violating ordinary canons of what makes sense. That, as noted just above, is nothing more nor less than exactly what he *means* to be doing. He does so deliberately, and presumably provocatively. We need to seek to understand (why).[2]

Lui is doing something perhaps similar to what poets do. He is knowingly saying something that ordinary language seems to rule out.

He says something that is (in a natural way of speaking) impossible to say.

I suggest that what he is doing bears close comparison actually to classic moments in the history of philosophy. To Plato's dialogues, in which the ordinary is often put into question, said by Socrates not to be good enough, questioned in the name of precision and exactitude. Or to Descartes's *Meditations*, in which the ordinary is bracketed entirely, in the name of rationality and a quest for certainty.

Lui is playing the role, seemingly a deeply felt, serious role, of the philosophical sceptic. The one who denies ordinary claims to know.

But in the case of Hiroshima, the quest is more specific than in the history of philosophical scepticism and thus not necessarily self-defeating in the way that such scepticism standardly is. Lui is not claiming that Elle doesn't know what love is, or courage, still less knowledge; still less that she perhaps doesn't know *anything*. He is questioning her specific knowledge *of Hiroshima* because of the special, novel character of what happened there. Something so terrible that the film seems repeatedly to suggest or imply that the local residents themselves are, in the main, quite keen to forget it too.[3]

One death is a tragedy, a million deaths a statistic. This remark, attributed to Stalin, seems to reflect a deep truth of human psychology. It can therefore be seen as framing the dilemma facing anyone wishing to make a film about Hiroshima (or indeed any cognate catastrophe). How can one possibly hope to approach its enormity, the endlessness of its horror?[4]

About halfway through the film, in the magnificent long (central) sequence when Elle goes deeper and deeper back into her memories of having lost her German lover to Resistance killing[5] during the war (this is nothing if not a film about what it means to "love thine enemy"),[6] Resnais takes us directly into that experience—giving us a vicarious sense of what she went through by a marvellous "trick" of filming. As we go deeper into Elle's memories, imperceptibly the sounds of the bar in which her "therapeutic" exchange with Lui is taking place fade further and further into the background, until they disappear entirely.

When Elle is brought harshly back into the present at the end of the sequence, suddenly the music "snaps" back up to full volume, and we notice that we have been accompanying her on her journey back into memory. That we have been with her in coming to terms with the trauma that she experienced. We have shared it so deeply we almost haven't noticed the depth of it. (Compare the way in which one journeys more with Folman—into his mindset and, just as in *Hiroshima*, into his memories—than one realizes, until the final portion of *Waltz With Bashir* forces one precisely so to realize.)

Perhaps the intensity of our "journey" with her is a good sign; perhaps it shows that we can come to understand what she is seeking to be understood in. Perhaps, if one can come to understand such a trauma, the tragedy of one death, *really* understand it, one might thereby take the first giant leap on a journey of 1,000 miles—the journey, in this case, of coming to gain some perception of what we (our leaders, our countries, here in "the West" where I write) did when we ushered in the era of atomic warfare.

The "therapeutic" journey that Lui leads Elle on here, and that we by extension are also led on, is one by means of which it becomes possible to emerge from the past, no longer in denial about it/trapped by it. One will be trapped if one gets stuck, in memory, or indeed psychopathologically, as in some depression (such as Elle evidently experienced). For the danger of grief, if it is "indulged" inappropriately, or misunderstood, is that it can descend into depression. Feelings of profound sadness, of one's world having a gaping hole in it,[7] can ossify into feelings of "permanent" negativity and aversiveness. But one will also get trapped by grief if one orients to it by way of denial, or by refusing to learn from it. Or simply by way of not really understanding the experience/emotion while in some superficial "ordinary" way imagining that one has understood it completely.

The film gives us a huge clue as to how we can escape the trap of history. A trap which, arguably, is present every time anyone says, for example, "75,000 people were killed at Hiroshima," and imagines that, just by virtue of stating this fact, they have brought light into the minds of their listeners, or indeed themselves.

For it turns out that if Lui is right, there is a sense in which Elle did in fact see nothing, understand nothing, know nothing, in/about Hiroshima. "How could I have avoided seeing it?" is one of the revealingly peculiar expressions she uses to characterize what *seems* the obvious truth: that she did see Hiroshima. But of course one can avoid seeing what one doesn't really want to see, or what is too grave to dare to see, by an effort of will (which may be unconscious).

Nevertheless she had the path to such knowledge, such understanding, within her, and thus potentially in her grasp, in a different form.

In a way, it's easy, to think or say what a sceptic does: that you can never know. Because it allows you to give up. It's harder to ask, What would it take, for you to be able to know? To be able to really see?

And thus we start to hear a deeper resonance in lines from the film such as "You can learn by looking carefully." Or "I begin to see"; and "I remember seeing before"—spoken by Elle as she recalls "therapeutically" to Lui her slow journey back to life as she recovered from the loss of her (German) love.

We begin to see.

***

How are we to understand the strange way in which *Hiroshima Mon Amour* ends?

What am I referring to? Superficially, the final portion of the film can seem much more straightforward than the opening epistemologically focused scenes. In the film's closing sequence, we follow the lovers as they prepare for the moment of Elle's departure, what will be their final separation. But there are two scenes which resist this apparent narratorial straightforwardness, and these two scenes cannot be ignored (especially because one of them is the very final scene, and thus has a good claim to be revealing the film's nature). They turn out to be keys to the film's philosophical, ethical and political meaning:

1. There is a scene a little while before the end of the film (the scene starting at 1.06.56) in which Lui is musing on the tragic fact that they, the lovers, will soon be separated forever. But then the scene takes a strange turn. In his absorption, from which he speaks aloud, Lui comes to think or to realize that there is a way that they might meet again: in war. If things go wrong, if the lessons of history (and of the film) are not learnt, perhaps it is possible they will meet again, only as enemies. (And/or as lovers? From "French meets German, to French meets Japanese," perhaps.) But this seems strange: because neither he nor Elle are soldiers. Elle is a woman at a time when women were not allowed into the military. And they are from opposite sides of the world (not as proximate as the German soldier was to Elle, in Nevers). And in a future war presumably Lui will be too old to be a soldier. So, again, what can Lui possibly mean? In a nutshell I suggest that it only makes sense for him to say they could meet again in war if it isn't so much him saying it as a person, but *as a place*. As the embodiment of Hiroshima, Japan. Perhaps Japan might meet the West again in war if the lessons of Hiroshima are not learnt, if militarism rises again, if we don't feel deeply the profound horror of what happened in World War II. The scene hints at, prepares for and is more fully explained by the ultimate and even stranger revelatory climax to the film:
2. In the very final scene a startling revelation comes to Elle, with laughter and wide-open eyes: that Lui IS indeed Hiroshima. And simultaneously, he sees that she IS Nevers, France. My interpretation of the scene is in a way straightforward, literal, but also quite radical.[8] We should see these final revelatory moments as the "characters" waking up to their actual role. Elle and Lui "wake up" to the fact that they are not, in the end, characters at all. Rather, they are small parts standing in for the wholes from which they come. They each represent their respective places, and

countries. It is those larger wholes, to which the film's viewers can be presumed more or less to belong, that ultimately matter.

I am suggesting that we don't really understand the film until its ending. *Until* these two scenes that round out the film, we appear to get further and further away from the mind-repelling reality of Hiroshima, and instead simply observe a sad and charming little love story. Yet the film is in fact about whether it is possible for us to understand something deeply unfamiliar, something we don't really want to understand (because it is too awful, and because in some way we may be implicated in it); in particular, the meaning of the appalling new moment in human history that was the dropping of the atomic bomb. What the film tries to stage for us is how hard it is to (bring oneself to) undertake this understanding; and how nevertheless, indirectly, it might be achieved. Perhaps France can come to understand the reality of the dropping of The Bomb by Western powers, not through going to a museum, nor through a fictionalized re-enactment, nor even through going to visit survivors in hospital, but in the first instance through/by coming to reflect more deeply on the awful reality of grief. On what it actually is/feels like to lose someone. By way of story, filmed intelligently. If we manage to understand what it is to lose someone close whom we care about—which is to lose an integral part of our world—then we might be able to attempt to understand the way our world changed when it became possible for tens of thousands of people to be obliterated in an instant. For *each of those people was part of the very world of someone else*. In much the same way.

Finally, we have an outline for a creative, human response to Stalin. Seventy-five thousand people died at Hiroshima: if we can actually start to *feel* the enormity of that, then perhaps we might learn from history.[9] Perhaps Hiroshima will be a nightmare from which we *can* awaken. Perhaps history will not repeat itself—perhaps The Bomb will never be dropped again—if, very roughly speaking, we can get enough people to experience Hiroshima *(mon amour)*. To have an understanding of the macrocosm through "experiencing" the microcosm and coming to see the connect between the two. To recall what it is to love someone,[10] which is perhaps best understood by knowing what it is to *lose* someone.[11] And to start to comprehend that to fire a nuclear weapon is to deeply warp the world of not just one, but tens or hundreds of thousands, or millions, of people. (I'm referring to the survivors and, crucially, their loved ones; let alone those who were obliterated. For every one of those obliterated or wounded people had a story, a setting.)

The end of the film then constitutes an alienation effect fully as radical, fully as philosophical (and, in its effects, political), as that which opened the film.[12] Symmetrically, the film closes with the viewer being told directly, this time by both the "characters," that the journey one took with them through the film was absolutely not just another film narrative. The psychical journey was to an altogether different destination, to a place where one is "forced" to be free, "forced" to seek wisdom. Placed in a position where it is harder not to start to understand what the previous couple of paragraphs encapsulate. Placed actually in, or *as*, a *place*.

One starts to be in a position properly to care about, to appreciate the preciousness of—or, as one might even put it, to love—Hiroshima. And, Oscar Wilde to the contrary notwithstanding, one will not kill what one truly loves.

The "characters," Elle and Lui, function as a kind of ladder we climb up, but which we then can—must—overcome.[13] By the film's end, their being characters is no longer a way, it is an obstruction. They had to appear to be characters in order for us to identify with them and to comprehend their grief. That was a necessary step, a major move in the indirect route to comprehension of the "incomprehensible" that the film proposes. But equally, they have to be characters no longer in order for us to accomplish the greater task that they have in the end facilitated: our understanding something far bigger than the little story of one or two or three human beings.[14]

Elle and Lui put us in a position where we can start to understand what it means when a place or people or nation savagely wounds another in an unprecedented manner. In order to complete the journey, we have to throw the ladder that they have offered away. We must take seriously the fact that this film is about France and Japan, the "West" and the "East," or indeed about any two "enemies," not (just, or even, in the end, at all) about a couple of individuals.

Hiroshima Mon Amour is, in this sense, a tremendous wake-up call for our times. These times are times when we desperately need to find a way to think ourselves back into the world-threatening seriousness of this being a nuclear age. Leaders exist now who are closer to using nuclear weapons deliberately than at perhaps any time since the 1960s. In particular, one Western and one Eastern.[15] And however, farcical this situation, it would, obviously, be profoundly tragic if it were to lead to war.

Hiroshima Mon Amour is in this sense an applied philosophical film just when we most need it. The film offers a deeply empathetic route to confronting the true horror of the ecological and human madness that would be the use of nuclear weapons. It suggests—and directly *facilitates*—an awakening to (and thus from) this madness.

And what should have become evident by now is that my reading makes sense too of the otherwise deeply puzzling title of the film. *Hiroshima, Mon Amour*; Hiroshima, my love. The macrocosm and the microcosm mixed together in one. Intimately related, as if they are themselves lovers. As if the film's title is itself saying to us that this is a film about how the macrocosm can be understood through the microcosm. I'm saying that it *is*. Love and grief point us towards understanding what it means to drop an atomic bomb (and understanding what such a bomb does)—and understanding what it would be therefore to determine never to do so again.

***

Once we've understood the film's end along the lines sketched above, then it becomes possible to develop a better understanding of Resnais's notorious and marvellous "puzzle" film that followed, *L'année dernière à Marienbad*. We can, I suggest, see the understanding developed above as a key to *Marienbad*. For if *Hiroshima* lacks characters in the true sense, but concerns a certain much larger movement of thought, a certain philosophical enlightenment, then probably—surely—*a fortiori Marienbad* does so too. My suggestion as to how *Hiroshima Mon Amour* works and what its nature really is makes good sense of the otherwise often weird acting in the film. A robotic weirdness, because the truth is too much. Or because Elle and Lui are not "in" their emotions, because of their traumas. Or, ultimately, because they're not real. They're not really characters at all.

All of which explanations are expanded upon in *Last Year in Marienbad* (*LYiM*).

My overview of the way in which the two films work as a brace would be this:

> *Hiroshima* centres upon *grief*. *Marienbad*, upon guilt, and/or regret, and/or *obsession*. These emotions are all much the same, inasmuch as they involve a kind of getting "stuck" in the past. But while grief makes us human, obsession risks making us merely egos run amuck.

Both films tell apparently a little story about a woman and her sequential two lovers. But more important, both focally concern something in the past that one doesn't *want* to face. (This not-wanting to face something is, according to Wittgenstein, *the* quintessential philosophical problem.[16] We tend to think that philosophical problems are intellectual problems; that they are solved by being clever. This *itself* is an exemplary running away from a truth: that the hardest issues in philosophy concern a lack of willpower—an unwillingness to face reality, to face others. And further, to face what we have done, to face the past or future, to face who we are, to face what conscience calls upon us to do or not to do.)[17]

Hiroshima enacts how one can come—*indirectly*—to understand something virtually incomprehensible, something so awful that the mind and soul rebel against the possibility of deep comprehension. An issue, or kind of situation, which one doesn't fully *want* to understand. It suggests that facts and figures will not be enough, and that standard narratorial devices or even pseudo-immersive experiences alone will not be enough. Rather, one needs to take a more indirect and empathetic route, a route that perhaps includes as a starting point an emotion one does understand or can (come to) understand: namely, what it actually means to grieve.

Marienbad is an image of the alternative to—the photographic negative of—this positive way forward. An image taken to the extreme. What

Marienbad offers is an image of rationality without empathy. Of culture without nature. Of the past as something that is systematically manipulated or denied or actively forgotten, rather than (as in *Hiroshima*) wisely sought after, reckoned with, learnt from. *Marienbad*'s "characters" ought in the end to be understood rather as adding up to a representation—an exemplification, indeed a symbolization—of a widespread and radically defective mode of reason, one fixated on representation and ratiocination. (A mode, moreover, that is thereby close to being philosophy's own default mode.)

Marienbad, as I see it, demands to be read as an allegory of our civilization, inasmuch as it shirks the kind of possibility offered by *Hiroshima* (and by *Gravity*; and by *Waltz With Bashir*; and in a sense by most of the films in this book). It is a nightmare of reason. And reason as a nightmare. Bringing to mind G.K. Chesterton's great epithet:[18] that the madman is not the man who has lost his reason, but rather the man who has lost everything *but* his reason. This, I suggest, is the meaning of the sterility of the chateau and the gardens; and above all of the deeply puzzling way in which the film ends. Why can't the (anti-)hero and heroine get away when they finally decide to flee, after the tall thin man who is the heroine's husband gives up on trying to stop them? Because they are stuck in a world where only the people, or (to be more precise) only their more or less Cartesian minds, are real, present. Thus the absence of anything but the people having true shadows, in one epochal scene in the film:

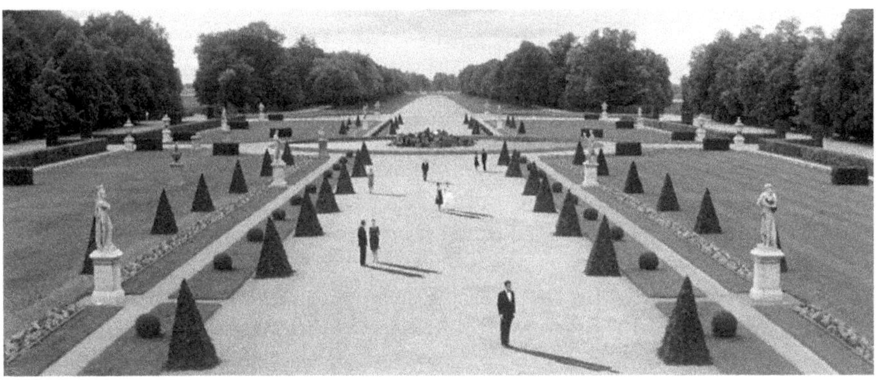

Figure 2.1 A life-world without nature, without materiality; without life, without world: *Last year in Marienbad.*

This is a nightmare of philosophy: a world of "Idealism." Because to be stuck in such a world (of thought alone) is to be stuck in a ghastly hall of mirrors.

Here is the tremendous final voiceover:

> The grounds of the hotel were symmetrically arranged without trees or flowers or plants of any kind. The gravel, the stone and the marble were

> spread in strict array in un-mysterious shapes. At first sight it seemed impossible to lose your way. At first sight . . . Along these stone paths and amidst these statues where you were already losing your way forever . . . Alone . . . with me.

". . . symmetrically arranged without trees or flowers or plants of any kind": Here one cannot help but think of *The Road*, of its world without plants, without *life*. But of course the strange thing about this point in the speech is that actually we see plenty of plants in the "sterile" hotel grounds of *LYiM*. It is as if those plants however have been devitalized by their subjection (in part, through extreme versions of pruning etc.) to the man-made world of straight lines and stones. Or, similarly, as if in the mind of the protagonist, even the life itself is already sterile, deader than dead. Forgotten, like the reality of last year.

In the mind of reason alone, be it ever so well arranged, there is no exit; no connectivity; no *groundedness*. And even when you are "with me" in it, you remain alone.[19]

Reason alone, within a culture that is culture alone, without nature, without a past that it is willing to discover the painful truth about, without empathy, reason that is narrowly mind alone . . . such reason is a nightmare from which there is no escape.[20] This life-world, this unworld rather, of nothing but constant haunting and repetitive thought, is the reality experienced by some sufferers of schizophrenia as documented by Louis Sass, and it is the reality often experienced, albeit not so extremely, by "normal people" in a culture dominated by scientism, by a humanism that risks occluding our animality and ecologicality,[21] by a derogation of the emotional, the empathetic, and the natural. (This latter fate, the fate of our culture at present, is brilliantly encapsulated by Iain McGilchrist in *The Master and his Emissary*. McGilchrist's account, at the very end of the book, of what the world would be like if it were dominated by the left hemisphere of the brain only, as he fears it is, bears striking similarities to the world as we find it in *Last Year in Marienbad*.)[22]

As the narrator says (in the passage quoted above) at the end of the film (and just this once, I think, we should believe him, unreliable though he is about everything to do with the past and about reality outside the confines of his mind—because what he is doing here is rather to allegorize his own condition), it should have been easy to escape from such a regularly patterned garden. But if you are stuck in your own mind, simply because you are unwilling or unable to leave it, then there is no escape. There *is* no outside.[23]

Hiroshima opens with a sceptical voice that is put to use: this is no idle hyperbolic scepticism of (say) a Cartesian kind. Lui is not denying that anything/everything exists; his denials are that ordinary experience, tools and/or facts etc. are enough to understand the enormity of Hiroshima. His responses to Elle serve to start to frame a route by means of which we might at last come to understand it adequately.

Whereas *Marienbad* dwells on the obverse case: a kind of scepticism in which the past is no longer something we can trust or rely upon, where the world itself is unreliable. Thus the narration in *Marienbad* is patently unreliable, and, more subtly, the chateau itself is unreliable: it "shape-shifts" at various points in the film. This is a metaphor for scepticism run amuck;[24] this is the danger of humanity getting stuck within the confines of itself,[25] trapped in mind, deprived of sane access to ecological placement, to the Other that is wild nature, or even to the Other that is other people. (Hell, here, is not so much other people as their unreliability or *absence*. As part of the absence, more generally and at least as crucially, of life, vitality, the non-us, nature, wildness.)

Those who do not understand the past are destined to repeat it: this idea one might see as being played with beautifully and disturbingly by both films. *Hiroshima* offers us the chance of escape from the trap of non-understanding, destination and repetition; *Marienbad* warns us more signally of what it is like when we get stuck in such a trap. Though the ultimate warning of course is already given in Elle's early words that form *Hiroshima Mon Amour*'s [*HMA*]'s apocalyptic challenge: understand Hiroshima, or nuclear war will happen again. (This time, presumably, immeasurably worse still.)

Roughly 75,000 people perished at Hiroshima. My argument has been that Resnais's film is a profound effort to enable us to understand that, as if for the first time. As something more like a tragedy than like a statistic.

Now let me start to move to conclusion, by re-comparing the two Resnais masterpieces in explicitly epistemological terms:

HMA is directed ultimately towards the world and confronts the difficulty of representing worldly disasters. It seeks to reduce this difficulty by way of an internal route. It shows how one can work through challenging microcosmic difficulties—which threatened to keep one stuck—and how this can yield a macrocosmic "pay-off."

Whereas *LYiM* is directed towards representations in a more "intransitive" way: it confronts the difficulty of escaping them, once one is focused on them and not the world. It explores how one *can* get stuck in the internal: it explores how getting thus stuck *is* a microcosmic disaster.

HMA opens, as laid out above, with a *specific* scepticism concerning the past. A scepticism not about memory in general but about a specific class of vast, mega-traumatic, extraordinary events, and whether they can be compassed within ordinary human experience of the kind that Elle has had in Hiroshima.

LYiM, as signalled by its obsession with representations (recall the use of the rococo; the most stylized of decorations where animal and vegetable are frozen into position in fixed, hard materials; the endless maps, and mirrors; the paintings, the plays; the repeated debate over the meaning of the sculpture of two lovers; the opening voiceover's

> obsession with mirrors, marble, stucco), is the world as representation only. (Combined with an unwavering will, that of the narrator—but a will not to truth, not to facing up to reality, but, apparently, to denying it.) *LYiM* seems to seek to generate a *general* scepticism about the past, about memory etc., but it is decidedly unclear whether such a general scepticism can be successfully generated; for the needful *contrast* between the ordinary and the extraordinary is not present. Just as the ordinary contrast between being alone and being together is problematically (would-be) eliminated in the final moments, especially, of *LYiM*, wherein it seems that the claim is being made that one is always alone, even when one is (in the ordinary sense) with another.[26] This is where Cartesian scepticism seems most *consequentially* to go wrong; is it something like this that is *demonstrated* in *LYiM*? Words are being used in ways that directly contradict their ordinary use/meaning; one wants, when one speaks in that kind of way ("I'm always alone"; "Perhaps only I exist"; "One can't *know* anything about the past" etc.) at one and the same time to retain one's ordinary language AND to use it in a completely extreme and novel way.[27]

So, if one can't actually *take* a general scepticism from *LYiM*, what *can* one take? What one can accept and understand is something true about a certain category or possibility of human experience: namely, devastating psychopathological experience as exemplified by "schiz spectrum" disorders.[28] So, although *LYiM* may not tell us anything about our actual or personal situation in life, it may tell us something about the situation in life of one who is subject to severe schizophreniform mental disorder (which it is unwise ever to assume as certainly not being a possible future experience one may have). And the intriguing thing is, the film suggests that such madness is of a piece with famous stances in philosophy: especially, with "subjective idealism" or solipsism. (Perhaps this accounts also for the insistent, grandiose, first-person voiceovers of *LYiM*. Is this the world of someone suffering a "schiz spectrum disorder" roughly as understood by Louis Sass, a world as if imagined and/or experienced to contain only me?)[29]

We should then venture that *LYiM*, while clearly based in trauma, concerns a response to such trauma that is psychotic, whereas *HMA*, while clearly concerned with the potentially unworlding effects of trauma, concerns chiefly neurosis and ordinary, though severe, emotionality.

LYiM concerns the nature of someone who has lost touch with reality: a person who, as it were, has taken seriously and got stuck in the kinds of arguments made by Descartes. What it is like to have lost touch with reality in this way is not to be in some exuberant manic world where one hallucinates things that are not there; it is to not feel as if *anything* is really quite there. Rather, for everything to feel as insubstantial as a play (as per the scenario of the opening and closing of *LYiM*).

HMA concerns the grief of Lui (et al.!) over Hiroshima—a national (and, in a way, global) loss—and the grief and depression of Elle over her personal loss, and the difficulty of finding a way beyond these emotions. The film *posits* a way through and beyond, in the indirect manner laid out above: by conjoining the two, or using one to get a handle on the other.

LYiM, dealing with the still more difficult case of psychosis, explores, as I've shown, what it is like if you literally can't find a way out, as is often the case for sufferers. Such people are stuck alone in their minds, like uber-serious philosophers; stuck, ungrounded, when focused entirely on—or contained entirely within—reason alone.[30]

Hiroshima Mon Amour offers us, *inter alia*, a philosophy of grief. It shows starkly how one loses one's bearings in the world when one loses *part* of one's very world, i.e. a person close to one. (Kelvin, in *Solaris*, directly compares grief over losing one that one loves to losing a limb).

And it shows how *hard* it is, in every sense, for one to remember, as time passes, as mourning occurs. As life goes on. The film in effect then submits that we all—if we have experienced or can comprehend such grief—have a route available through which we can do what we *ought* to do: that is, to lose our bearings a little, or a lot, in the face of the ultimate horror, nuclear annihilation.[31] The challenge to which Lui and Elle rise (and which the protagonists of *LYiM* cannot rise to) is one of daring to remember, to actually face the past; and thus face the present and also the future, without illusion. Those who refuse to remember the past are doomed to repeat it. (Thus, as Lui puts it in the latter part of the film, "I shall think of this story as of the horror of forgetting.")

This is of course the challenge to which Japan and the world as a whole must rise. Forgetting, failing to learn, will expose us to the risk of the past repeating itself. A risk that, in today's nuclear world, simply cannot be countenanced.

The kind of commitment that I see *Hiroshima, Mon Amour* as asking for viewers to enter into, evidently, is not narrowly intellectual. It is not just about solving a puzzle—the puzzle of this strange and difficult film. Neither is the yet more puzzling puzzle that is *Last Year in Marienbad*. One won't come to . . . *dissolve* any such puzzle[32] unless one actually *feels* the horror of the dead end of cold rationality that it stages—and thus critiques. And in this sense my line of thought means to be compatible with what Robbe-Grillet says in the Preface to the screenplay for *Last Year in Marienbad*.[33] In the end, thinking of this film as a puzzle to be solved *is exactly to stay in the trap that the film itself sets*.[34] Rather, one must stop being confined within thinking alone. What is required is an emotional *involvement*; even an involvement, one might put it, of your *soul*. The film asks you to really *experience*, to enter into and not pull back from or refuse the difficulty of facing what is on the screen. It also asks that you do not passively observe what occurs merely as a stream of images. You are invited *to remember* and make real your experience, in terms of what you think/do next. *HMA* manifests an awakening. *LYiM* concerns the alternative: it's about being determined to remain asleep, stuck stagnant or dead.

Resnais asks us to commit in and from these films. One might say: the *meaning* of the films is not a matter solely of mere intellection. His aim is much more important. The films' "meaning(s)" includes *the actions we, as viewers/an audience undertake(s) in response to the films.* We are asked to change our lives; to live differently, now we have seen them. And the asking doesn't stop.

Hiroshima, Mon Amour keeps *demanding* of us, every time we think of it, let alone re-watch it, or each time something profound arises in our lives (or our polities) in terms of the choices that we make which are relevant to the concerns the film dramatizes.

In this way, it is like most great films. Such as those explored by the rest of this book.[35]

Notes

1. For such an understanding, including of this would-be repulsion of understanding, see my analysis of grief in "Can there be a logic of grief?," in Kuusela et al. (eds.), Wittgenstein and Phenomenology (London: Routledge, 2018), www.researchgate.net/publication/324831961_Can_There_be_a_Logic_of_Grief. See also Matthew Ratcliffe's work on grief.
2. This effort to understand can be paralleled to Wittgensteinian efforts to understand what Heidegger means, for instance in speaking of angst, or of how "the nothing itself noths." See for instance Ed Witherspoon's PhD thesis, *Nonsense, Logic and Skepticism*, https://philpapers.org/rec/WITNLA.
3. In this regard, Lui's stance resembles Claude Lanzmann's stance in relation to what he claims to be the unrepresentability, except through certain indirect means, of the Holocaust. See his epochal *Shoah*. (Cf. also n.4.)
4. One might compare here Resnais's documentary *Nuit et bruillard*, where similar questions are raised, concerning the topic of that film: the Holocaust.
5. She was alone in hating the liberation, as Hiroshima was alone in hating the war's end. If we can really come to understand this "from the inside," we are achieving the task set in *Hiroshima Mon Amour*.
6. In this way, as we shall see later in this book, it bears an intriguing similarity to the *Lord of the Rings* films. See especially n.61 and n.62 in Chapter 6, and *supra*.
7. For my picture of grief as logically and phenomenologically having this character, see my "What is grief?: a personal and philosophical answer", https://medium.com/@GreenRupertRead/what-is-grief-a-personal-and-philosophical-answer-d83d7f288c96, and my "Can there be a logic of grief?". (See also the discussion of Stone and her dead daughter in *Gravity* in Chapter 5.)
8. And therefore I cannot agree with Hunter Vaughan's judgement, at p. 125 of Where Film Meets Philosophy: Godard, Resnais and Experiments in Cinematic Thinking (New York: Columbia, 2013), that "*HMA* ends by nullifying its experimental philosophy and returning to an order of meaning founded on traditional rules of thinking." There is little traditional in recasting persons entirely as places, in really moving from the micro- to the macrocosm.
9. Elle clearly feels bad about what happened to Hiroshima, while Lui thinks of her as having the wrong kind of feeling, a feeling with the wrong kind of cause. (Is it, perhaps, that she is *blocked* from having a genuine feeling about Hiroshima—or

anything else?—until she has come to terms with her feelings for/from Nevers? I think so. [Thanks to Jerry Goodenough for this insight.])

10. We need to understand, to *feel*, that and how the human macrocosm comes down to *second-person* relations. (Whereas statistics, curiously, mostly refer to individuals, not to families, friendships, personal stories. Thanks to Mihai Ometita for emphasizing this point to me).
11. As *Gravity* will have it (see Chapter 5): launching is landing. Leaving is arriving. To understand what it is to love, you must understand what it is to lose.
12. And in this way the end of the film can be seen as completely radical a shift as the ends of the two films examined in the previous chapter (and as the ends of the films examined in Chapter 4, and of *2001: A Space Odyssey,* as examined in Chapter 5). (Cf. also n.8, above.)
13. Compare the discussion of *Waltz With Bashir* in the previous chapter—and the authorial strategy of Wittgenstein, most strikingly in his Tractatus.
14. Cf. also my remarks about how this point helps us to frame successfully the otherwise peculiar acting in Resnais's films. (The acting is sometimes a kind of alienation effect, just as the start and end of *Hiroshima* are.)
15. And, more generally, our world is replete with militarism (including in Japan) and nuclear weapons poised to launch (including in France). Bear in mind here that while the film was being made in 1958–9, France was not quite yet itself an atomic power; but de Gaulle was making the plans for the nuclear *force de frappe* that would emerge in 1960.
16. See Lecture XI of his Lectures on the Foundations of Mathematics, *PI*, p. 109, and the "Philosophy" section of the Big Typescript (London: John Wiley, 2012).
17. On this point, the (Wittgenstein-influenced) work of Hannes Nykanen is very instructive; see his The 'I', the 'You' and the Soul. An Ethics of Conscience, Doctoral Dissertation (Åbo Akademi University Press, 2002).
18. From Orthodoxy (London: Simon & Brown, 2012 (1908)).
19. I return to develop this point below. It may seem a pessimistic reading of the final words of this final speech; but I think it has to be the correct one, in the context of the film, where the two central characters never really meet at all, and walk out for that final "journey" automatistically, and not even side by side or hand in hand.
20. We miss the real danger that Cartesianism's contemporaries saw in it, if we think purely in terms of dualism and the non-physical mind, or even in terms of scepticism. For, for contemporaries of Descartes, the real problem was with the world left behind if the mental and the spiritual are extracted and re-located in some non-physical realm. The world of our lives then becomes an austere and unfeeling world, a world to be described purely in mathematical terms of number and extension, to be totally describable in terms of the new physical sciences. A world of extreme regularity and without life. A world with remarkable resemblance to that of *LYiM*. . . . (Thanks to Jerry Goodenough for the inspiration behind this note.)
21. Cf. David Ehrenfeld, The Arrogance of Humanism (Oxford: OUP, 1978).
22. For my take on McGilchrist's book, go to http://users.skynet.be/tony.aerts/images2/About_TheMasterAndHisEmissary_IainMcGilchrist.pdf and scroll down. (For an alternative vision of what a society NOT dominated by compulsive thinking would look like, see e.g. Eckhart Tolle's work.)
23. Compare *PI* 103, which examines just such an inclination.
24. The film's implied critique of this scepticism, this unreliability, shares form with a central thread in Wittgenstein's anti-'private-language' considerations. When there is no public shared world, then there is nothing to rely on, and everything might as well be constantly changing. See e.g. *PI* 258 and 293. Moreover, trying to "fix" things (as the narrator of *LYiM* is constantly seeking to do), within this

ungrounded fluidity, only makes things worse. In the wrong context, the desire for clarity makes things worse. Nothing is clear, in this world: not even whether the meeting that allegedly took place between the lovers took place at *Marienbad* or somewhere else (see the scene at 27.49, where we learn that this too is unclear). In this way, the film's title is itself a kind of wonderfully deliberate fallacy of misplaced concreteness, of over-stated clarity.

25. I would argue that humanism, and anthropocentrism, reveal just such stuckness. See e.g. my "A price for everything?: The 'natural capital controversy.'" Co-authored with Molly Scott Cato, *Journal of Human Rights and the Environment*, 5: 2 (2016), 153–167.
26. How problematical this kind of move is brought out in the final chapters of Sass's Madness and Modernism.
27. Stanley Cavell is the greatest analyst of this kind of problem. See his work from J.L. Austin and from Wittgenstein, on this, in The Claim of Reason (Oxford: OUP, 1979).
28. As Louis Sass analyzes them. (For exposition, see Part 2 of my Applying Wittgenstein (London: Bloomsbury, 2007).)
29. See Sass's book The Paradoxes of Delusion (Ithaca NY: Cornell, 1994) for why such solipsistic imagination, paradoxically, turns out to seem to require the existence of others. . . . Cf. also my "On approaching schizophrenia via Wittgenstein" Philosophical Psychology 14:4 (2001), 499–514 (www.academia.edu/207833/On_approaching_schizophrenia_via_Wittgenstein).
30. These issues will re-occur when we get to *Melancholia*, later in this book, which concerns neurosis (depression; a kind of grieving in advance, a pre-emptive sadness)—and possibly also, on another more violent and extreme reading of the film, world-catastrophe in Sass's sense? (*Solaris*, another great film to compare with *Melancholia*, as I shall do, appears to concern especially psychosis, though also, very clearly—indeed, starkly—grief. Thus all four films can be seen to tie into each other—as I'll sketch, in the Conclusion to the present work.)
31. These issues will return to us in the chapter on *Gravity*, which also considers seriously grief and the limits of our thinking about it.
32. For that is more like what one needs to do: not solve the puzzle, but let it dissolve, loose its grip on one, and give way to a different mode of comportment that does not trap one in analytical thought.
33. New York: Grove Press, 1962.
34. Here there is a partial connect of my line of thought with that of Impossible Puzzle Films, by Miklós Kiss and Steven Willemsen (Edinburgh: Edinburgh University Press, 2017).
35. Big thanks to Naomi Marghaleet of You Said It Ltd. for vital editorial suggestions in this chapter. Thanks also to Mihai Ometita, Julian Hanich and Jerry Goodenough for helpful comments.

3 Learning From Conceptually Impossible Versions of Our World

Never Let Me Go (and *The Road*)

This chapter, more than most others in the book, explicitly reflects on the broadly Wittgensteinian way of taking film/literature-as-philosophy that this book practices: it explores the nature of liberatory, *transformative* Wittgensteinian film-as-philosophy. It does so of course via a case-study/ example-led approach: a case study principally in the Romanek movie / Ishiguro novel *Never Let Me Go [NLMG]*.

This essay proceeds in this manner because this IS a central part of what adequately represents the theme of "therapeutic"/Wittgensteinian philosophy: proceeding *via examples*. Because, almost uniquely, and as previewed in the Introduction, above, Wittgensteinian philosophy deliberately does not provide a theory of film/literature, no matter of what kind. Unlike most philosophers, we Wittgensteinians don't believe in "theory." We figure practice as what counts, logically. (Thus our emphasis on "ordinary language" etc.) Wittgensteinian film/art-philosophers believe that films/artworks *themselves*—those films that can successfully be seen, roughly, *as* works of therapeutic philosophy—constitute this practice, "in collaboration" with their audiences. There is no further task for the theorist / the philosophical critic to undertake from a position of would-be superiority.

I emphasize this here because all this will be particularly clearly visible in *Never Let Me Go* (and *The Road*). As I shall intimate, one needs only to see these films with open eyes, to see the philosophical work that they do. But that work is work that they do on one, with one, through one. Through many in fact. Through *you*, us; through the involved "audience."

I will argue that part of the way to understand *Never Let Me Go* is as what Wittgenstein, in the *Philosophical Investigations*, calls *an object of comparison*:

> 130: Our clear and simple language-games are not preliminary studies for a future regimentation—as it were, first approximations, ignoring friction and air resistance. Rather, the language-games stand there as *objects of comparison* which, through similarities and dissimilarities, are meant to throw light on features of our language.

> 131: For we can avoid unfairness or vacuity in our assertions only by presenting the model as what it is, as an object of comparison—as a sort of yardstick; not as a preconception to which reality *must* correspond. (The dogmatism into which we fall so easily in doing philosophy.)

One learns (as an audience-member/reader) from *NLMG*'s differences from our actual world, as well as from its similarities.

The filmic situation is also like some of Wittgenstein's imaginary scenarios, such as the world's weirdest "woodsellers," seemingly described in his *Remarks on the Foundations of Mathematics [RFM]*:[1]

> People pile up logs and sell them, the piles are measured with a ruler, the measurements of length, breadth, and height multiplied together, and what comes out is the number of pence which have to be asked and given. They do not know "why" it happens like this; they simply do it like this: that is how it is done. . . . Very well; but what if they piled the timber in heaps of arbitrary, varying height and then sold it at a price proportionate to the area covered by the piles? And what if they even justified this with the words: "Of course, if you buy more timber you must pay more"? . . . How could I show them that—as I should say—you don't really buy more wood if you buy a pile covering a bigger area?—I should, for instance, take a pile which was small by their ideas and, by laying the logs around, change it into a "big" one. This might convince them—but perhaps they would say: "Yes, now it's a lot of wood and costs more"—and that would be the end of the matter.—We should presumably say in this case: they simply do not mean the same by "a lot of wood" and "a little wood" as we do; and they have a different system of payment from us.

According to the "resolute" (therapeutic) reading of Wittgenstein of which I am a proponent, the point about this scenario is that, as we *try* to imagine it, we learn something about the limits of sense. For there is nothing which counts simply as succeeding in imagining these people simply as selling wood as we do (as Wittgenstein remarks in the final sentence). "That" is a conceptual impossibility.

We try to imagine "the woodsellers," and the limits of our success tell us something about the "limits" of our world.

Similarly, *NLMG* is an exercise in pushing a boat out further than it can actually be pushed. In watching this film/reading this book, one leaves the bounds of sense. I mean that quite literally—I argue here that the "society" shown in the work is conceptually impossible. Remember that clones are identical copies, which only have a different upbringing. The principal reason for holding the society of the film to be impossible is that the clones, the "students," not only do not rebel, they have no thought of rebelling. This environmentally produced totalized subalternity, as I shall explain further below (and in Chapter 6), is outside the realm of human possibility.

As in Wittgenstein's "philosophy of nonsense," this film allows us to look back in at our world from an impossible outside, and thus learn something about sense and about ourselves/our lived world.[2]

Never Let Me Go's breaching of the limits of plausibility and, more important, of the very limits of sense (a pushing found in a number of other major philosophical films: including especially, as I shall explain, *The Road*; and also, at times, the films of Alain Resnais, as explored in the previous chapter; and, perhaps, the two films that are the subject of Chapter 4) is available in the service of increasing one's/our intellectual autonomy. Of *liberating* one/us.

When the ability to question the facticity of one's existence is put into question (as is the case *within NLMG*), *we* immediately have the capacity to *answer* that question differently. *Never Let Me Go*, in that way, *enacts* film/literature as philosophical therapy as liberation: for it creates, in anyone willing to listen to it, willing to *see* (it), a philosophical capacity, and thereby a political capability; one that is lacking in all of the protagonists *in* the work. I mean of course *really* willing to see it, or willing to really see it, in something like the *Avatarian* sense of those words (see Chapter 6, below); the problem here, once more, is (and once more following Wittgenstein) a problem of the will, not primarily of the intellect:

> What makes a subject difficult to understand—if it is significant, important—is not that some special instruction about abstruse things is necessary to understand it. Rather it is the contrast between the understanding of the subject and what most people *want* to see. Because of this the very things that are most obvious can become the most difficult to understand. *What has to be overcome is not difficulty of the intellect but of the will.* [*Nicht eine Schwierigkeit des Verstandes, sondern des Willens ist zu überwinden.*][3]

So, if you really do have the will, let's go on.

***

How does this remarkable work, *Never Let Me Go*, work? First, how does it have the terrible effect that it has on one?

I think it is a lot to do with one looking and looking for *hope* in the thing, and each time being thoroughly frustrated. And with being forced to imagine people doing this to each other. And moreover: that they (some of them, namely the "clones") do it to themselves, or let it be done to themselves. That you do it to yourselves.

The frustration of this: at there being *no* rebellion. Worse: they have NO *thought* even of truly running away.

This is what makes one feel so very sad, or indeed *sick*. If at some point in watching this film you don't feel a little sick (and not from the gore and the surgery scenes, which are in fact extremely mild, almost veiled, as perhaps we want them to be), then I suspect you haven't really seen it. You probably need to watch it again. You're in denial.

Feeling sick at this film at times is an *appropriate* reaction, one without which there is no understanding, because of the link in cases such as this between comprehension and affective reaction. Someone who took *Never Let Me Go* purely intellectually would be a dangerously cold person. They would be, in fact, the kind of person who might conduce to the lived possibility of creating a society alarmingly like that featured in *Never Let Me Go*.

Then again, denial is to some degree an appropriate reaction too. The scenario is so . . . *unthinkable* that to be too easily able seemingly to think it would (again) suggest too great a facility with the kind of horrors gradually unveiled in this work. Too great an ability not to be horrified. A certain amount of denial turns out to be a *mode* of eventual acceptance and of taking seriously,[4] for it is a mode of acknowledgement of the utter wrongness of what is displayed before one, its unworlding horror and (thankfully) impossibility.

One is not just sick when confronted by the crunch moments, as the protagonists' plight fairly remorselessly worsens. (Even after Ruth's death, the closest we ever get to what is needed is Kathy and Tommy girding their loins to go and ask for a "deferral."[5] They *ask* their oppressors politely for a few years' grace.) Part of the effect is created in the most "delightfully" frustrating way: in the film, and even more so in the book (because it is longer, more filled out with chit-chat etc.), one comes increasingly not to be able to bear the way that the central characters spend so much of their time engaged in gossip, in playing and engaging in romance, in small talk, in worrying about and playing out their friendships. Anything, seemingly, to avoid having to face the nature of their existential and ethical and political reality. They casually play out their time in this way,[6] rather than thinking of any kind of escape or rebellion. One wants to shake them, to wake them.

It becomes truly unbearable.

[But in this, perhaps they mirror us? For isn't this what *we* mostly do too: gossip and fiddle, while the world burns.]

Of course, at first, one can't judge them too much for this; they are children after all, and children need to be allowed their innocence, and their gradual finding out about what is really important and how things really work. But such judgement *builds* during the film. With what might be characterized as their refusal to become real adults, to achieve *any* autonomy from their "guardians' " vision for them.

There is a great line early in *Never Let Me Go*, in those years of innocence: "Who would make up stories as horrible as that?," asked of Miss Lucy, the would-be "decent"[7] attempting-truth-telling teacher, by one of the children, of the scare-stories about what would happen to schoolchildren who left the school grounds.[8] Of course, there are some things that really cannot be told: being told is not possible. Being told is still not being told. This is why the moment when one sees the donor whose eye has been taken and who Kath is caring for is so appalling. For, while one has already been told by this point very clearly of the nature of this world, to actually start to . . . *see* its consequences—to *look them in the eye*—remains a shock. This is telling: we've been half-blind, willfully unseeing. We think we have accepted the facticity of this terrible world; but finding it so gut-wrenching and dismaying to see this young deliberately one-eyed woman, and to see her and Kath *simply* accepting the fact of it, tells us that in fact as yet we have not.[9] If one is warned explicitly about *Never Let Me Go* before watching/reading it (as I was), it is still I think impossible to imagine just how awful an experience it is going to be, to watch it. It cannot be believed, that people could make up stories as horrible as the story that it actually is.

There couldn't be any such society as this.[10] As already laid out, because the absence of even any thought of resistance or rebellion could not be taught into one. And also because there simply couldn't be humans that lived together like *this*. If something like this were going to be created, it would (I would suggest) have to involve for instance a more rigorous apartheid between the "humans" and the "subhumans" (the name that I shall sometimes use for the clones in *Never Let Me Go*). And also stronger propaganda against the latter.

Everyone in the film treats the "subhumans" as subhuman even though they are literally indistinguishable (except for some social gaucheness etc.) from the "humans."[11] And, at a level of fundamentals, of course, they are. They lack more than the most superficial basis for differentiation and prejudice; they lack even the difference of "race" or (thinking of *Blade Runner*) of replicanthood. They are *exactly* the same (they are clones; that is what makes them such perfect "donors," perfect matches).

The film depicts then a *society* that could not quite be. The real issue, as I shall explore below, is how much reassurance we can take from that.

***

Never Let Me Go is a *mirror image* of *The Road*.[12] That the scenario in *The Road* (of the entire biosphere—except a few non-natural companion animals—dying but some human beings remaining alive) is presumably biologically/ecologically impossible is often seen as a weakness of *The Road*. I regard it, instead, as a *strength*, for three reasons:

1. It tends to focus attention and interest where the film (and book) requires attention primarily to be placed: on *the human beings* who occupy the scenario.[13] It asks us to reflect honestly on whether what is depicted in the film is *humanly* possible, and to consider the alarming likelihood that it *is* / would be.[14] Philosophically speaking, it focuses our attention on the Heideggerian/Arendtian distinction between *world* and *Earth*.[15] It also gives us a remarkable insight into the non-permanence of the Gibsonian "affordances" of *our* world. In *The Road*, what is afforded (by objects, by people etc.) is often radically different (from what *we* are used to), or less.

2. More important, it *requires* us to be clear in the end that it is impossible (in more ways than one) for us, humans, to survive without a functioning ecosystem/biosphere.[16] Thus after focusing on the human beings (1), we get even clearer on their (our) utterly embedded/dependent nature (2). We need to recognize our *utter* tiedness to the biosphere, our being nothing without it. The only scenario upon which we can imagine human beings surviving without nature is an impossible one.[17]
3. Perhaps most importantly (and most subtly) of all, that this artistic work depicts a scenario that is biologically etc. impossible makes it easier to go into denial about it. It gives one an "out."[18] This might be thought a weakness, but is in the end a deep strength. For it *mirrors our reality*. This "out" enables one to go into denial about the alarming potentialities of human nature under situations of extreme stress; and it enables one to go into denial about the likelihood of a future quite like shown in *The Road* being actualized. We take shelter[19] in the literal falsity of the film's premise—and thus hide from its metaphorical truth, and its closeness to literal truth. This indirectly gives us the opportunity to recognize ourselves and our situation better. For our actual situation—of being on a road that, unless departed from, will bring us to a catastrophe not that dissimilar to *The Road*'s envisaged future—is precisely one in which we look for any excuse in which to hide in such denial. The deepest truth in *The Road* is this way that it mirrors our own desires back to us. It seems to offer us something startlingly other and scary; but one fairly swiftly finds reassurance in its excessive premise. *This* is exactly how we are condemning ourselves or our children to an utterly disastrous future, by always looking for such "outs." *The Road* is at its deepest as a "therapeutic" or "liberatory" work, in encouraging one finally to take the difficult step of recognizing *this*, (3): recognizing, and then *overcoming*, the temptation not to act, the temptation to remain in denial, by means of "cleverly" dismissing the film as "not realistic."

A parallel mirror image argument—changing as it were left to right but otherwise leaving the essence intact—can be made for *Never Let Me Go*. Its physical/biological *plausibility* (on this front, it is science fiction, but could easily become science fact) dovetails intriguingly with what in the key respect I have started to adumbrate above is its utter implausibility humanly speaking. As I've said, the most striking manifestation of the latter—what, thankfully, makes the film definitively impossible, conceptually impossible, for human beings—is the total absence of rebellion, of even the *thought* of rebellion, among the "subhumans."

I find that many people who read or hear what I have to say on this film resist this point. They mournfully argue that it could be fully possible. I don't think that they are reflecting deeply enough on the conditions of possibility for the human animal, which exclude a total defeatedness,[20] a total absence of autonomy, of ability to at least imagine acting otherwise. That ability surely just comes along for free with a sufficient level of intelligent thought, of language, of sociality. If one can imagine history, for example, or if one can imagine anything being different, or if one can simply imagine (or indeed experience) genuine anger, anger directed at an object, then one can to some degree at least imagine rebellion.

The very fact that someone finds it disturbingly easy to think that their world could be our world, and is still *disturbed* by thinking this, itself SHOWS THAT WE ARE NOT IN THAT WORLD. And that is an encouraging thought.

But this not reflecting deeply enough on this crucial point is *itself* very revealing. It is I believe a way in which we protect ourselves: from the *pain* involved in realizing that we can never be completely beaten, the pain that comes from having to continue to *try*. We can never have an excuse for not carrying on fighting; and that's hard.

We are looking to be able to give up the struggle against the "overwhelming" odds that something not *altogether* unlike the kind of profound/impossible societal failure depicted in this film and other films I am mentioning in the present chapter—the kind of failure likely to both result in and result from dangerous climate change, for instance—may well occur. We are looking for what Sartre called a "reprieve." We know how heartless and quasi-fascistic and how near to death our current society is; and thus the overwhelming temptation to wallow in negativity, to give up resisting. We then resist admitting what I am saying here: that we should recognize that a society like that in *Never Let Me Go* could not be. This resistance itself can teach us; our inclination to think that the impossible could be actual is a clue: to seeing clearly how bad our situation is, and to seeing clearly our vulnerability to the siren call of "reprieves" from having to act. But we must go further: to *overcome* this resistance, this inclination, so that we can practice actual resistance to the mega-machine.

Never Let Me Go fantasizes a total absence of even the *thought* of escape. This is, one might say, a society in which a very special scientific fascism has achieved total victory, such that there is not even what Orwell called "thoughtcrime." This is not something which can make any sense, as a vision of human beings.[21] There couldn't be humans, society, where there is no *dream* even of rebellion or escape. The children are trained to obey, to fear escape, to accept their role; and (in the film version) there are constraints such as the bracelets they wear that keep track of their returnings home; but these are in no way adequate to explain their entire sheeplikeness, their going *more* placidly and willingly than lambs (let alone pigs or dolphins) to the slaughter.

Some think that *The Road* and *Never Let Me Go* are bad science fiction, because of their (complementary) impossibilities.[22] This is a shallow definition of what makes for good sci-fi. For it leaves out precisely the factors that I have enumerated above, which are precisely what makes these films/books into, I would suggest, very great sci-fi. Into "therapeutic," politically motivating, transformative philosophical works. Works that generate what Wittgenstein calls (*PI* 127) *reminders* of basic things that we forget because they are always before our eyes, or because we don't *want* to be reminded of them. (Compare here *NLMG* p. 259: "The world didn't want to be reminded of how the donation programme worked.")

Never Let Me Go is also in this regard reminiscent of *Blade Runner*.[23] For example: The replicants of that film have a four-year life-span; the life-spans of the "subhumans" are similarly fixed, though not quite as absolutely and not to so short a length. In a way, *Never Let Me Go* is *Blade Runner* in *reverse*: there is absolutely no rebellion, no coming to person-status of any of the "creatures," and no one *in* the film to *see* them.[24]

"You poor *creatures*,"[25] says Madame, at the end.[26] The film could be well seen as a metaphor for the normalization for this way in which we *do* treat non-human animals.[27] There probably *will* be non-human animals bred to provide us with organs, soon. Note that the post-Hailsham "farms" for raising these donor-children are characterized, chillingly, as akin to "battery-farms."

Tamed and domesticated animals are bred *so as* to be non-rebellious. And this makes it easier not to have to acknowledge them adequately; because there will be no "come-back" from our failure to do so.

This, the work's absence of mutual *acknowledgement*,[28] is why it is so depressing, so negative, so very very sad. But unlike *Blade Runner*, it is *not possible*: you can breed sheep etc. to be . . . like lambs, but the conceptual capacities that make it possible to understand the nature of—the necessarily non-total nature of—authority,[29] to have some understanding of history, and so forth, cannot be dissociated from the capacity to disagree, and therefore the imaginability of resistance.

It is a slightly consoling thought: that what we are shown in *NLMG* is not literally possible, for humans.

The fact that the world this work represents really is worse than our world (so bad, that it is impossible) gives a little comfort. Perhaps we can even take some minute *pride* in that fact? Not really—because one can't take pride in not doing the impossible. (This point yields a contrapositive of Kant's famous "Ought implies can": "Can't implies irrelevant to ought.") And because most of what is in the film is *metaphorically* real in our world.

What do I mean by this?

Well, the speech from the Charlotte Rampling character, Miss Emily, near the end is very important, in at least two ways:

- Firstly, we are taken down and taken aback by the revelation that Hailsham was concerned with *ethics*; that people like her and "Madame" are the closest that the "subhumans" had to *advocates*;[30] that they were trying to show to sceptical others that the clones had souls (at all) with the gallery etc.; that Hailsham was as good as any such place has ever been, and better than it is now. This is a completely unexpected, disturbing revelation. But perhaps it shouldn't have been. It is akin for instance to the phenomenon of the "nice" concentration-camp guard, or to how good Rudolf Hess or Mussolini can look if you place them alongside Hitler.[31] (For us, it raises a disturbing possibility, perhaps the awful truth: maybe we should be grateful to be living now, mired though the world seemingly is in an eternal Treblinka for animals, in an ecology veering from bad to worse etc., simply because maybe this is as good as it gets?)

- Secondly, and more crucially still, take the Rampling character's (Miss Emily's) brilliant, chilling remark that it is pointless to "ask people to return to darkness"; "People are never going to return to the days of lung cancer, motor-neurone disease."[32] The scenario of the film should I think be read among other things as a metaphor for consumerism and materialism, for the ratcheting-up effect of expectations and the "normalcy bias," for growth, for an inability to contemplate making the big changes and "sacrifices" that need to be made to prevent eco-catastrophe.[33]

When watching this film, the hard thing to do is to allow oneself to fully feel it. Not to give in to the pull to escape from *it* in an escapist way. It is as if the film asks you: never let me go. Or: never let me let you let go (of me). It is hard, to rise to this challenge. It requires an effort of the will.

I have a not-infrequent desire to scream like Tommy (though usually at people rather than at the night sky, and, as I discuss below, that is actually an important difference).[34] I want to wail out my rage, horror, I want to shock others at last into being awake at what we are doing to our home and ourselves. To quote William Faulkner, about a somewhat similar wail, that some would say is full of sound and fury, signifying something awesomely terrible, a bellowing not from Tommy but from (Faulkner's) Benjy: "[There was] more than astonishment in it, it was horror; shock; agony eyeless, tongueless"; "It might have been all time and injustice and sorrow become vocal for an instant by a conjunction of planets."[35]

Tommy is terribly important; like all the actors in the film,[36] Andrew Garfield does a magnificent job at portraying him. I think overall he / Tommy is actually the star (both as adult and as child), overshadowing perhaps even his more famous female co-stars. There is something on the edge about him, *throughout* the film: he is a kind of icon of the insanity of the system; he is a living, walking symptom. He is full of sound and fury, so full that he is never at ease.

He *is* the symptom. He is a microcosm of the whole.

For the thing about his bellowing is that it seems somehow to lack an intentional object in the way that anger or rage ought to have one. He bellows this tragic fury because he seems to have no sense any more than the others do that anything could actually change. His fury is real and extreme; but it is the fury of a victim, not an agent.

Tommy's strangeness throughout is a brilliant counterpart to Carey Mulligan's character's (Kathy's) loveable care and normality—which ultimately risks being a kind of lived deadness. It is as if she isn't quite *alive*, precisely because she is well balanced during all of this complete insanity. He fails to see reality: his mad fantasies about the art gallery; his saying "It's weird, but it's a good weird, isn't it?" as they three are going to see the boat together, when in an obvious way "it" is very far from a good weird; . . . but yet at some level he KNOWS.[37] On the surface, Tommy, who seemingly doesn't see reality, *symptomatizes* the society both as it likes to see itself (for most of the time) and as it is really, in his unconscious, which erupts in his bellowing. Kathy sees reality more easily; but she somehow still doesn't KNOW. She doesn't really manage to or allow herself to feel it. In the end, it is she rather than he who is more in denial. In her stoicism (so typical of Ishiguro's characters, subject to his sympathy but also his searching critique),[38] she is inhuman, impossible, incomplete.

In our world, we are saying: "We can live without the coral reefs;[39] without the Amazon rainforest; without most cetacean cultures;[40] without the passenger pigeon, without the Tasmanian tiger, without the tiger"; and on and on; *we are as a whole living without our organs*. The Earth is gradually becoming the unpleasant truth of what a "body without organs" would look like. Though this is using the term rather differently from Deleuze and Guattari: "The Earth," they write, "is a body without organs. This body without organs is permeated by unformed, unstable matters, by flows in all directions, by free intensities or nomadic singularities, by mad or transitory particles."[41] That is, we usually think of the world as composed of relatively stable entities ("bodies," beings); but these bodies are "really" composed of sets of flows moving at various speeds (rocks and mountains as very slow-moving flows; living things as flows of biological material through developmental systems; language as flows of information, words etc.). My sense of the term "body without organs," drawn from Ishiguro and from this film directed by Romanek, is obviously more simple, not positively valenced as Deleuze and Guattari's is: their Earth as a "body without organs" is the earth allowed to flow; but when you remove its real organs, its "green lungs" etc.,[42] as we as a species are doing, then that flow is broken up, and it starts to collapse. So I think actually that what I am saying is, although first appearances might have suggested quite otherwise, on reflection entirely compatible with what Deleuze and Guattari are saying, if I understand them aright.

As later Heidegger enables us to understand,[43] the very concept of "resources" is problematic. *Never Let Me Go* shows the horror of seeing humans as resources, as inherently potentially replaceable, harvestable. But at a deeper level, this film and *The Road* can be read as showing too the horror of taking animals or indeed Nature as resources.[44]

***

I have been suggesting that at the level *of mind and society* the tale *Never Let Me Go* tells is literally impossible, indeed *conceptually* impossible. At the level of the world *of sentient beings* (i.e. recalling that the majority of that world is constituted by non-human animals), however, it is more or less literally true. We have domesticated most of the world;[45] and we treat nearly all non-human animals nearly all the time as if we live precisely in the *NLMG* world. We do treat livestock (and many wild animals, too) as abhorrently as the "clones" in the film are treated.

Similarly, at the level of the living world as a whole (the world as described for instance in Joanna Macy's *World as Lover, World as Self*), the conceit of *NLMG* (the centrality of organ 'donation') is more or less metaphorically true. (Or, Macy and other deep ecologists would say, literally true; think Gaia.) We are like people who gradually remove bits of our own lungs, because they make pretty trinkets. Our forests etc. as functional ecosystems *are* these "green lungs."

At the level of our world as a whole, the work describes (us) fairly accurately—and it is in that way alarmingly directly *alike* to *The Road.*

This is terrifying, repugnant; and it *forces a response.*

Now, I've argued that it is not true that there could be a society just like that depicted in the film of *NLMG*, where children were farmed to become adults whose organs could be harvested and where those young adults could basically interact freely with the general populace (the bracelets that in the film version they have to wear are the *only*—and weak—direct preventer of flight or fight) yet in which no one rebelled *nor even thought of rebelling.* It is just not true that there could be a society like that, no more than that there could be a (non-)biosphere with humans remaining alive in it as depicted in *The Road. But* it *is* true that right now our world as a whole, in terms of its overall dynamic, in terms of the balance of governing forces, is alarmingly similar to the world of these two films.

The gravity of the eco-crisis etc. is such that it can seem that *for all the good that our "rebellion" etc. thus far is doing, we might as well be in the world of the film.* We are donating away our living plant's vital organs.[46] *Willingly.* And, as things spiral downwards further, it will be the children who suffer most of all.[47] (Actually, thankfully, the thought the first sentence emphasized just above is not necessarily true: some things are changing for the better;[48] but reflecting on how alarmingly close to being true it is what film can help do. It can enrage us, and get us to *see*—and *feel*—what needs doing.)

The truly terrible thing is that at the very end of *NLMG*, Tommy and Kath let each other go, without even trying to fight to hold onto each other, without even *considering* trying to do so. This contravenes every convention of movie-making, every urge in our minds, every particle of humanity. This makes figural the ways in and the occasions on which we do the same. The sense in which we are profoundly at risk of letting our world and our children go.

One of the most awesomely depressing moments in the film is very close to the end, when we re-watch the scene of the last eye contact between Kathy and Tommy, as he goes to what will be his death on the operating table. It is so wordlessly heart-rending, that they have let each other go,[49] and that Kathy is now left alone, the boat on the beach. The appallingness continues, with Kathy's words after the screen has faded to black: "It's been two weeks since I lost him." The language here is delicate. "Lost" is a euphemism, as all the distinctive language of the work is.

Ishiguro's writing of this on the final page of the book is masterful.[50] He/Kathy continues,

> I was thinking about the rubbish, the flapping plastic in the branches, the shore-line of odd stuff caught along this fencing, and I half-closed my eyes and imagined this was the spot where every*thing* I'd ever lost since my childhood had washed up, and I was now standing here in front of it, and if I waited long enough, a tiny figure would appear on the horizon across the field, and gradually get larger until I'd see it was Tommy. (Italics added)

Look at the use of the term "everything" in this sentence, where "everyone and everything" or some such would be more appropriate . . . unless one has tacitly absorbed the prejudice that renders the clones less than fully human. Subtly thing-like. Repositories, "standing reserves" of organs waiting to be harvested. (The way we typically actually do think of most non-human animals and of the planet.)

The rubbish can be seen as the detritus of our consumptive society, a key sign of what's gone wrong in our relation to Nature. Amidst all the garbage which we are gradually turning the Earth into, it isn't surprising if one gets tempted to turn humans too into stuff, things.[51]

Possibly the most devastating moment in Miss Emily's climactic revelatory speech is that (already indexed earlier) when she remarks, "There was no way to reverse the process. How can you ask a world that has come to regard cancer as curable, how can you ask such a world to put away that cure, to go back to the dark days? There was no going back."[52] This speech sums up our predicament. The issue applies *a fortiori* to our lack of care for non-human animals, and for our descendants (*vis-à-vis* climate apocalypse etc.). The challenge is somehow to find a way of getting the mass of humanity to give up things—cheap flesh to consume, cheap consumer crap and cheap car-fuel etc.—that they (we) have completely accustomed to, and which have been thoroughly normalized. The challenge is, if one says this, to not be instantly dubbed as wanting to go back to "the dark ages," as (for example) Green politicians often are.

The challenge is to find an effective way to rebel against and overturn this vast short-termistically self-interested hegemony. The challenge can only be risen to by starting to get people who don't want to do so to *see* in a different way.

In willing the "donors" to rebel, in exhausting ourselves doing this, *what we are really doing is urging ourselves to rebel.* Our urging and our urgency is thus perhaps ready to be put to the test, into action.

Realizing how bad things are and how in such a situation one simply must resist is quite parallel in some ways for instance to how *Avatar* works on one.[53] But these characters cannot even dream for themselves, for a better future. So: we have to dream (and act) for them.

They (*especially* Kathy) pull themselves back into themselves at every turn. They avoid care, they avoid love, to avoid the pain of vulnerability.[54] They let one another go, pre-emptively. A communal response to oppression, political activity, is *not possible* for atomized individuals who have withdrawn into silos. This makes a sense of the book's bizarre impossible scenario. System change is impossible for quasi-solipsists.

So, this is another transformative, therapeutic, liberatory work. A work to enlighten us, in lieu of them having an "us" to potentially awaken.

And here is the importance of the great "Buddhist"/"Heideggerian"/ "Tractarian"/*Blade Runner*-ian moment at the very end of the film (only), a wonderful *addition* to the ideas present in Ishiguro,[55] in Carey Mulligan's character's (Kathy's) very final words[56]—which I found myself working up to during the latter half of the film independently, the first time I watched it—of realizing that life is always only in the present[57] (and thus that there really is an important respect in which the "humans" are no better off than the "subhumans," and perhaps worse off in not realizing how desperately they are in denial—about their own mortality; half-imagining that, with enough transplants, they can live forever).[58] This is a final crucial way in which this artwork exposes to the light our normal denial about death.[59]

Here is that final speech:

> What I'm not sure about is if our lives have been so different from the lives of the people we've saved. We all complete.[60] Maybe none of us really understand what we've lived through . . . or feel we've had enough time.[61]

Just like *Blade Runner*, this work explores a world in which *no one* is truly human.[62] For it is not only that you can't be truly human when you are as oppressed / as deprived of personhood as the "students" are. The *oppressors* thereby deprive themselves of real personhood, real humanity, too. In their failure to *acknowledge*, they mutilate *themselves*. (They gain the "subhumans" organs, but cut out their own souls, in the process.) The interesting question, then, is whether *we* are really any different, at the macro level at least. (Think of our relations to future people. We are cutting out/causing to fail their organs, both metaphorically and [at a temporal distance] literally, for our own temporary "well-being.")

This film is then very like *Blade Runner*; only the scenario is darker, much worse. For these simply are human beings, who are simply not treated as human. The film puts this more severely, in Miss Emily's final speech, than the book does. Charlotte Rampling's character says that the gallery was to prove, in a beautiful phrase damning with faint praise, that the clones were "all but human." Garland's screenplay/Romanek's film also does a masterful job of subtly demonstrating the "untouchability"[63] of these "subhumans"; the *only* time in the film that they are touched by a non-medicalizing "human" hand[64] is the culminatory moment that I noted earlier, when "Madame" touches Kathy's cheek at the end—and says, "You poor *creatures*" (emphasis added).

Religion (or even just anything that we would recognize as a genuine ethics) is strikingly absent from the film's world.[65] There is no more "re-ligare,"[66] reconnection, binding (in unity, with the divine), acting in the assumption of full interconnectedness. As I emphasized above, in putting my line of thought alongside Deleuze and Guattari's, the parts of the whole are far more expendable when seen as separate/separable. Expendable organs. The "subhumans" are not God's creatures. They are man's creatures (they are clones; the excuse, presumably, for them being treated differently from their "originals"). The sense of sacredness to *creation* has gone.

They get, roughly, at best, only the kind of "care" and stewardship that non-human animals are "entitled" to, in our world as it is: i.e. hardly any at all.[67]

Never Let Me Go helps you (question your own ability to) question the facticity of your existence: it helps bring into focus the extent to which your "facticity" is ideologically convened. But such questioning is itself of course then immediately available in the service of increasing one's/our intellectual autonomy.

Such liberation is not easy; but it *is* available. But it really isn't easy. My guess is that most people watching the film (or reading the book) massively go into denial about it, while they are watching as well as afterward. This denial might take various forms: simply resting content with the thought that "it couldn't happen here"; simply finding the failure to rebel of the main protagonists somewhat pitiful a failure and thus blaming the victims; perhaps also tacitly identifying with the "humans" in the society depicted; failing to see wide lessons for our own lives, or metaphorical meanings of the work; failing to feel deeply with the characters; *simply* being "depressed" by the whole thing—that last response I would bet is very widespread, and (if so then) that is a deep failure.

However, one natural way of reading this book/film gains its plausibility and coherence by resolutely sticking, despite the temptation not to, to the perspective of the main characters—and then trying to find *analogues* for those in our own world. I have earlier emphasized the aspect of conceptual impossibility. But as the chapter has gone on, I've emphasized the analogues aspect more and more. These two aspects can, I would hazard, sit alongside each other, because the concept of an "object of comparison" explicitly allows for seeking after and learning from both similarities and differences.

This observation in turn brings out something deep about liberatory philosophy: that the nonsense is to an extent perspectival (the perspective is afforded by what Wittgenstein calls our "form of life,"[68] as well as, more specifically, by one's purposes etc.). For I have written here about the extent to which the film is implausible to the point of impossibility. I have indeed called it even a conceptual impossibility: these are not human beings; this is not a society. (And that is not just an "ethical" or "political" remark.)

But what if this is both? Nonsense on one view, but rendered sensical from another perspective.

It's like with the *Tractatus*, or indeed the *Philosophical Investigations*, properly understood.[69] The film is simply nonsense, from one point of view, *at one moment in the dialectic*. It doesn't add up to a human possibility of any kind. From another point of view, *at another moment*, it is revealing a whole raft of literal and metaphorical human possibilities, (possibilities) which, typically, *we do not want to see*.

This is one reason why it is crucial to compare and contrast Miss Emily's final speech with Tommy's scream. From the inside, as it were, there is no possibility of articulating the metalogic to which the main protagonists are subject; all one knows/feels is rage, at most (and those that feel this much are the exception). The rage is manifest in a scream. The rest of this society seemingly see all this as normality and thus acceptable and not something that can be wound back/reeled in. And those who do see it for what it is—*us*—can articulate the crime, what's at stake etc.; only from the outside.

We seek to console ourselves perhaps by noting that the crime is too complete to be even possible (humanly). Or, in relation to *The Road*, that the scenario is too extreme to be even possible (ecologically). Though at this point we should note another way that the two films differ. Remarkably, despite its incomparable seeming-bleakness, *The Road*, with its redemptive ending, and with its central metaphor of the fire that the boy must keep—morality, the one thing worth saving from the old society, and living on in a boy who is not of the old society, thus giving hope of something new arising that is worthwhile—is *less* dark than *Never Let Me Go*. The boy is a new Prometheus, snatching a *symbolic* fire (*not* the technology that has sent us to ruin)[70] from the godless ruins of our world. *But* even though *Never Let Me Go*, like *Last Year in Marienbad*, is, by contrast, uncompromisingly dark, it nevertheless offers *us* some very real light. It en-lightens.

And this brings us to be able to reflect on the way in which these films at multiple levels play with the concept of what can be imagined: of characters who are not imaginative enough, of what can be learnt by seeking to imagine the unimaginable (including a society where what is unimaginable exceeds *what* is actually intelligibly unimaginable), and of our own (overactive or) under-active imaginations.

One looked and looked for hope in *NLMG*, and didn't find much, or perhaps not any at all. But it turns out one was looking in the wrong place. The glimmer of hope in the present chapter, a glimmer that I hope to embroider on and enlarge, quite explicitly, in Chapters 4–6, is elsewhere. But very "close to home" indeed. The place to look is within oneself and one's fellow cinema-goers, when the lights go up; where one knows one may be able to find it. How does one know this? One knows it *just* because *one was looking for it* (*in* the film),[71] *so very hard*. (And because one found the film so very hard to take.) And what *matters* then is how we decide to move forward from this viewing; whether we decide to allow the world at one level or another to become more and more like that of this film—or not. The argument of this chapter has been, as one might put it, about how an ideal viewer/reader would respond to *Never Let Me Go* (and *The Road*). Part of the answer is clearly: by not remaining content to be a viewer/spectator, but by *acting* radically, where the characters in the work entirely failed to—and failed even to imagine the possibility of doing so. By giving up an essentially spectatorial relationship to life. The kind of relationship typically encouraged by movie-going[72]—and by philosophizing, so long as philosophizing itself remains spectatorial,[73] and not, as it should be, "therapeutic," *liberatory*. And so long as movie-going is conceived of as escapism rather than as a step on the path to wise action. Agency.

So, who would make up stories as horrible as *Never Let Me Go* and *The Road*? Answer: Ones who wanted us to end our dogmatic, complacent or despairing defeated slumber. Both stories concern adults who tell children "noble" lies.[74] They raise starkly the troubling question of what we ought to tell our children, at a time when their very future is being radically compromised.[75] The only way to avoid such a predicament without evasion is: to change the future.

One might even risk saying[76] that artists have too often largely only interpreted the world; the point, as any *true* philosopher or filmmaker will realize, is to change it.[77]

Notes

1. RFM I-143ff. (Cambridge, MA: MIT, 1978). The way that I see the "woodsellers" pseudo-case as working is modelled on that of Alice Crary (in "Wittgenstein and political philosophy," in our The New Wittgenstein [TNW]), and David Cerbone (in his "How to do things with wood," also in TNW), which in turn are founded on Stanley Cavell's (in The Claim of Reason, especially pp. 115–125). For my own twist on the "case," see Chapter 6 of my A Wittgensteinian Way with Paradoxes (Lanham, MD: Rowman & Littlefield, Lexington, 2012).
2. I described this process already in outline in my Introduction to Film as Philosophy. It is precisely modelled on a central method of Wittgenstein's, *engaging* with our tendency to utter nonsense, finding a (deviant) kind of "*use*" for it. As in the instance of the "woodsellers," quoted above.
3. Emphasis added. "Philosophy" (chapters 86–93 of the Big Typescript), p. 161. Corresponding to TS 213, Kapitel 86.
4. See https://medium.com/@GreenRupertRead/what-is-grief-a-personal-and-philosophical-answer-d83d7f288c96 for more background on what exactly this means.
5. However, I will return to an important point that emerges at the very end of the film, and that of course we have all known all along (but that we are perhaps

typically in denial about): What do any of us ever get, in terms of seeking to prolong our lives, except a few years' grace?

6. This feeds the poignancy, the terrible sense of regret, at the close of the film. Tommy and Kathy should have seized their moment much earlier. (See the remarks towards the end of this chapter on living in the moment, as dwelt on beautifully by Kathy at the close of the film.) Their not doing so could be seen as of a piece with their letting each other go (to their deaths: see the discussion of this later in this chapter). Which in turn brings to mind the protagonist of Coetzee's Disgrace (London: Vintage, 2000), at the very end of the book, letting his favourite dog go.
7. See pp. 79–80 of the novel (Croydon: Faber and Faber, 2010 (2005)).
8. This is very reminiscent of course of much the same thing done to Truman in *The Truman Show*. The parallels don't stop there; like Truman, the kids in *Never Let Me Go* are lied to from the very start; they are born into a false world. Unlike Truman, they are then gradually told the truth. Or rather, as Miss Lucy puts it: "told, and not told."
9. Cf. also Ishiguro's delicate (tacitly self-reflexive) characterization on pp. 36–37.
10. This is a stronger claim than the ethico-political claim/suggestion, that I also think correct, that liberal (or Thatcherite) society is not really a *society*.
11. Their fashion sense is a little odd or old, a little bohemian. Their clothes are handmade (or cast-offs); their "school uniform," on closer inspection, is all individual. Ironically, while they are supposedly "copies," they are, even as kids seemingly with a school uniform, more individual and individuated in their apparel than "normals" are in many English schools. . . .
12. Taken together, they offer a kind of "completeness".
13. Think here of Primo Levi's great autobiographical works, discussed in Chapter 6. His fearful inquiry into whether those who survive, when things get sufficiently bad, are never the best of us. That rather, the drowned were the best of us.
14. Though one should be very cautious before assuming that what appears to be McCarthy's somewhat Hobbesian/negative view of human nature and its potentialities is *true*: On this point, see especially Rebecca Solnit's hugely important book, A Paradise Built in Hell: The Extraordinary Communities That Arise in Disaster (London: Penguin, 2009). Cf. also my "On preparing for the great gift of community that climate disasters can give us," *Global Discourse* 7 (2017)—Issue 1: "After Sustainability—What?", pp. 149–167.
15. See e.g. p. 32 of the book (London: Pindar, 2007): "On this road there are no god spoke men. They are gone and I am left and they have taken with them the world." The world is gone; only the Earth remains.
16. I explore this point more in Chapter 5.
17. This may help with the deeply difficult task of being honest with ourselves, and with the next generation, about what is happening. See e.g. www.ueapolitics.org/2016/10/17/rupert-read-addresses-ueas-new-students-with-a-shock-message/. *The Road* considers this difficulty partly by the difficulty the man has in knowing how honest (or otherwise) to be with the boy. At times, he takes a "noble lie" stance: see e.g. pp. 106–7 of the book.
18. And we desperately want such an "out." Not least because the work's scenario is so hard to imagine wanting to live in. We perhaps think, Beckett-like, "I can't go on." I can't bear to go on reading/watching/*feeling* this. In such a reality, I wouldn't be able to go on (see *The Road*, p. 145). But we should remember what comes next, in Beckett; namely: "I'll go on." Thus the extraordinary power of "life goes on" in *The Road*, and the extraordinary fact of the book's (and, perhaps more so still, the film's) having a "redemptive" ending: see close to the close of this chapter.
19. And here one ought to think of deep meaning in (the title of) the film *Take Shelter*, as discussed shortly, in Chapter 4.

20. I examine this version of the point in detail in Chapter 6, via *Lord of the Rings*.
21. I shan't enter here into the further question of when something like this might become a marker of being genuinely "post-human." Is genetically enabled fascism possible in a post-human society, a new species, yet one similar enough to us for point-by-point comparison to still seem conceptually possible? When do we stop being human? How can we decide what are the criteria for behavioural patterns that are beyond the human? *Blade Runner 2049* begins an intelligent investigation of such questions; it suggests that it is challenging to get to any point of being in such a way genuinely post-human while preserving intelligence, motivation and any kind of operational independence. (Which tends to support my line of thinking in relation to *NLMG*: that the scenario it imagines falls apart under its own weight / is conceptually incoherent. *A fortiori* to *Blade Runner 2049*, clones, identical copies of humans, would have to have at absolute minimum something like the capacity for rebellion that we see gradually emerging in *Blade Runner 2049*.) How deeply challenging it is to imagine a genetically enabled "post-human" fascism in which a *total* docility is bred into human animals is in fact one of the main reasons why I shan't enter into the "further question" here; for perhaps it isn't really a very well-formulated question yet. (Thanks to Sergio Fava for the idea of this note being needed.)
22. One looks back in from the outside of our world, and has learnt something about sense and about oneself, or about one's lived/social world. *That* outside, however, is impossible. It does not—cannot—exist. (Cf. Philosophical Investigations 103.)
23. On which, see Stephen Mulhall's great writing: e.g. the relevant chapter of the first edition of On Film. Cf. also Jeremy Goodenough and Rupert Read, "Review: Timothy Shanahan (2014), Philosophy and Blade Runner," in *Film-Philosophy* (2015), https://ueaeprints.uea.ac.uk/57018/.
24. In *Avatar*'s sense of that term. Beyond what LYiM calls what one sees "at first sight".
25. See especially p. 267 of Never Let Me Go. On pp. 256–57, Miss Emily refers repeatedly to the students as having been "reared." Madame adds immediately, "I wish I could help you." This is one of many subtle moments of denial in the film half-hidden behind some of the unsubtler denial (e.g. as generally practiced in the school, Hailsham) that blankets much of it. For of course, if she really wanted to, she would help them (e.g. by hiding them in her cellar, or in a thousand other possible ways).
26. And what is it that Tommy creates his marvellous drawings etc. of? Non-human animals . . .
27. Compare the brutally satirical *South Park* episode "Whale whores". (The kicker is in the final line of that programme.)
28. See Cavell's and Mulhall's work on this. We might helpfully put the point thus: The students are looking for a *criterion* for (their own) humanity. For a way that they (and others) can *know* that they are human. But exactly the demand for such a criterion is a *sign* of inhumanity. Instead, one should take up the second-person stance, and (mutually) acknowledge.
29. This inevitability is a main topic of Chapter 6, wherein I set out how *Lord of the Rings* shows us that, while we may *want* to believe in total subalternity, that desire is itself a running away from our inevitable agency.
30. This is splendidly laid out in some of the posters that the *NLMG* DVD lays out for us from the "Hailsham Campaign": e.g. "Are you moved by the plight of our donors? Concerned about the living conditions of our donors? Then join us. // The Hailsham Campaign is at the forefront of a plan to change our donors' way of life. Its intention is to give you the chance to receive organs that come from INDIVIDUALS that have had a DECENT life." Comment is superfluous. This text makes nice connections with Miss Lucy's concept of "decency", and her

out-of-line remarks accordingly to the "subhuman" children while at school; and with the metaphorical presence of non-human animals in the film, rather casting doubt perhaps on (e.g.) campaigns to "free" battery hens by placing them in "free-range" pens etc. Cf. e.g. Chas Newkey-Burden, "Free range is a con", www.theguardian.com/commentisfree/2017/jan/30/free-range-eggs-con-ethical.

31. Or in Amazon's *The Man in the High Castle*: the remarkable feat of casting *Hitler* as the moderate, compared to Heydrich et al.
32. Of course, the "subhumans" are by implication figured by this remark as *not* people. (I explore this speech at greater length supra to n.52.)
33. Growthism is a tacit form of denial about death—see my article "Growth and death", here: http://oneworldcolumn.blogspot.com/2010/07/growth-and-death.html. This film forces one to face one's/our denial about death. Though some viewers will be so appalled that they will probably retreat deeper into denial; I come back to this thought before the end of this chapter. (This thought should admittedly make one ponder somewhat about how *effective* the film is likely to be as a politicizing therapy.)
34. I am not alone here; for instance, Caroline Lucas MP has on a number of occasions confessed to a similar level of frustration.
35. Quotes from the final part of The Sound and the Fury (New York: Vintage (the corrected text), 1991 (1929). The source text of course is these troubling and symptomatic words of MacBeth's, from Act 5 Scene 5 of MacBeth, "Life is a tale told by an idiot, full of sound and fury, signifying nothing." MacBeth thinks life has betrayed him. Such actual betrayal we witness in *Never Let Me Go*, but, appallingly, it is only us who adequately witness it, not any of the characters *in* the novel/film. Ishiguro's intertextual referencing of Faulkner (and thereby of MacBeth) seems pretty plain on p. 270 for instance, where, after his despairing agonizing bellowing in the night after their deferral request has been turned down, Tommy says to Kath: "I'm sorry about just now, Kath. I really am. I'm a real *idiot*." On p. 269, his face was described during this episode as contorted with "*fury*"; back on p. 9, in the key childhood episode of bellowing, he is indeed described as "*bellowing*"—the term invariably used by others of Benjy, the "idiot" in The Sound and the Fury.
36. Including Keira Knightley, as Ruth, who is wonderfully, appositely annoying at the Cottages. Finally, a role tailor-made for this actress. . . .
37. As Kathy concedes herself, and Tommy, on reflection, allows, on p. 270 of the book: in fact, Tommy's unconscious knowledge is probably more present in the book version than in the film.
38. See my review (published in *MIND* 112:447 (2003), pp. 506–9) of Anthony Cunningham's brilliant treatment of Ishiguro's critique of Stoicism and of Kantian "duty" in *The Remains of the Day*, in his The Heart of What Matters, http://rupertread.fastmail.co.uk/heart%20of%20what%20matters.pdf.
39. See e.g. "Great Barrier Reef Imperilled as Heat Worsens Die-offs", Jacqueline Williams, www.nytimes.com/2018/07/04/world/australia/great-barrier-reef.html.
40. On which, see my "How whales and dolphins can teach us to be less stupid", www.opendemocracy.net/transformation/rupert-read/how-whales-and-dolphins-can-teach-us-to-be-less-stupid.
41. A Thousand Plateaus (Minneapolis: U. Minnesota Press, 1993 (1980), p. 40.
42. See "Forests—Earth's lungs", http://blog.ontrac.com/bid/63022/forests-earth-s-lungs.
43. In The Question Concerning Technology (New York: Harper Perennial, 1977 (1954), especially.
44. This way of taking Nature has become alarmingly widespread. See e.g. my joint paper with Molly Scott Cato, " 'A price for everything?': The 'natural capital controversy'," *Journal of Human Rights and the Environment* 5:2 (2014), 153–67.

45. See www.ynharari.com/topic/ecology/—wildlife is now only a small percentage of life on Earth. Of course, this fact doesn't mean that we have really succeeded in domesticating *the planet*, as some ideologues of "the Anthropocene" like to fantasize. On the contrary, as I am exploring (or simply noting) in this chapter and in other parts of this book, our interference with the planetary ecosystem is increasingly coming back to bite us. An instance of potentially catastrophic significance is the plunge in insect numbers: see Damian Carrington, "Warning of 'ecological armageddon' after dramatic plunge in insect numbers", www.theguardian.com/environment/2017/oct/18/warning-of-ecological-armageddon-after-dramatic-plunge-in-insect-numbers.
46. See my "Some thoughts on civilizational succession" and "Climate change: once we no longer deny it then we just might have the will to drastically change course", at www.truthandpower.com/rupert-read-some-thoughts-on-civilisational-succession and www.thelondoneconomic.com/opinion/climate-change-once-we-no-longer-deny-it-then-we-just-might-have-the-will-to-try-drastically-to-change-course/14/03/, for support for this claim.
47. As well as non-human animals, of course, once again underlining how good a metaphor *NLMG* is for their plight (a plight literalized to the extreme in *The Road*, where they have been eliminated). Consider for instance the astounding, silencing fact that humanity has eliminated half of all the world's wildlife, over the last 40 or so years: see "Half of world's wildlife lost, says new WWF report", www.worldwildlife.org/press-releases/half-of-global-wildlife-lost-says-new-wwf-report.
48. Consider for instance the recent upsurge in awareness of the plastics crisis engulfing the world, an awareness most strongly influenced by the BBC TV series *Blue Planet*. This is leading to some surprisingly bold responses; see e.g. Namaan Zhou, "All single-use plastics should be banned by 2023 [Australian] Senate Inquiry recommends", www.theguardian.com/environment/2018/jun/26/recycling-senate-inquiry-recommends-all-single-use-plastics-be-banned.

    Or consider the dramatic rise in veganism.
49. Cf. NLMG, pp. 118–19. They are *supposed* to let each other go. "At the Cottages . . .when a couple were saying goodbye to each other, there'd be hardly any words, never mind embraces or kisses." (Cf. here the great, terrible episode of Cowslip's warren, Chapters 12–17 of Watership Down (London: Penguin, 1971).)
50. As is Mulligan's acting in the film version, as she stands barely holding it together, staring both emotionally and stoically into the distance from where no one will come running.
51. As nature is being turned into "capital," commodified (cf. n.44, above), now the same is starting to be done to humans and communities: see https://twitter.com/aledjones_gsi/status/986178568718831616.
52. NLMG, p. 257.
53. As I argue in the chapter on *Avatar*.
54. In this way, they "foreshadow" for us the understanding of depression expertly offered in *Melancholia*: see Chapter 4. Cf. also my discussions of Denethor and Theoden in Chapter 6 (each of whom could be read as a little akin to King Lear as understood by Stanley Cavell in his masterly reading of that play, in "The avoidance of love," in Must We Mean What We Say?).
55. Though possibly fully in line with what he intended. Otherwise, the remarks he made to the *Guardian* about the book would verge on the utterly bizarre, rather then merely being somewhat (ahem) in tension with what obviously emerges from the story that he actually wrote: " 'I thought, certainly at the planning stage, it was my most cheerful book,' he says, smiling. 'Unless you have a real sense of precious things under threat there would be nothing sad about time

being limited. The people in the novel believe, irrationally, like we all believe, that love can do all kinds of things that make you exempt from your fate.' " (Ishiguro, speaking in an interview with the *Guardian*: www.theguardian.com/books/2005/feb/20/fiction.kazuoishiguro). I challenged Ishiguro on this point in person, recently, at UEA, and he responded drily that one has to take into account how *very* uncheerful his books are.

56. Though again, as always, the Carey Mulligan narratorial voice is unacceptable in its *acceptance* of the world which the "humans," and not she and her kind, have made.
57. And once again, this acceptance of our life-span is just what is needed, if we are to overcome the trap of egoism and the danger of using up and throwing away the resources needed by the future for the sake of attempting to "prolong" the present, prolong our individual existences. As dramatized for instance in Joss Whedon's *Dollhouse*. Such acceptance is needed, and would be a genuine alternative to the fascistic mentality hegemonic in *Never Let Me Go*; but the paradox of course is that such a mentality would/will be hegemonic forever if *it* is accepted, not rebelled against.
58. As well as being in some denial of course about what they are complicit in.
59. There is a connection hereabouts with Malick's *The New World*: both films are about how to love life while being in the midst of something structurally utterly intolerable. America is a gigantic graveyard of the Native Americans. Americans are happy beneficiaries of a holocaust. Similarly, the "humans" in *Never Let Me Go*.
60. The concept of "completion" is *NLMG*'s euphemism for death. In the case of the clones, it has of course a certain disturbing literalness to it; their mission is complete, when they have done what they were designed for: been harvested.
61. And cf. these much more ambiguous words of Tommy's, both true and false, at p. 277 of the novel, after he has "recovered" from his final bellowing episode: "[I]n the end we can't stay together forever."
62. Thus my employment of scare quotes around the term "humans" turns out to have been peculiarly appropriate.
63. They are "taboo."
64. The perversion of medicine in the film is of course of interest in itself. Particularly powerful, I think, is the scene in which a nurse speaks to Kathy, using all the standard nurse tropes, but in the cause of this miraculously perverse non-caring system of treating the "subhumans" as patients—but patients whom one is not seeking the long-term wellness of. (Also: presumably the "subhumans" work as "*carers*" precisely so that the "humans" don't have to. Don't have to *feel* and suffer and get soul-tired as the carers do.)
65. Compare (and contrast) *The Road*. And *Lord of The Rings*—see Chapter 6.
66. www.etymonline.com/word/religion Thanks to Sergio Fava for this point.
67. Cf. the appearances of animals in *Waltz With Bashir*, as discussed in Chapter 1.
68. See e.g. PI 240–42.
69. Understood, that is, according to the "resolute" reading of those works; see my and Crary's The New Wittgenstein.
70. Cf. my discussion of *2001: a space odyssey* in relation to technology, in the closing portion of Chapter 5.
71. Our complicity perhaps in Ruth's need to be loved; our complicity surely in the boy Tommy's need to be accepted; and above all our complicity in Carey Mulligan's (Kathy's) character's gossip and outsiderness and loveliness—these are telling. They are what need to be overcome by on the one hand what, on an (overly) optimistic reading of the film's end, the Carey Mulligan character ultimately perhaps achieves: an understanding of what it is to be alive, to see, to

experience the moment, not to get caught up in anxiety about prolonging one's existence; but also by, on the other hand, what is totally absent in the film, and what *you* must supply: a spirit of rebellion, a political awareness and sensibility, a refusal to accept the utterly unacceptable.

72. See e.g. Walker Percy's philosophical masterpiece of fiction, The Moviegoer (New York: Vintage, 1961).
73. On which, see especially John Dewey's later work, especially The Quest for Certainty (Delhi: Isha, 2913 (1929).
74. See especially p. 263 of Never Let Me Go.
75. And at a time when we are exterminating wild animals and torturing domesticated animals in unprecedented numbers etc., as noted above. See my own direct engagement with this question, *vis-à-vis* actual living students, here: https://medium.com/@GreenRupertRead/why-i-had-to-tell-my-students-that-i-fear-for-them-64bf1625b878.
76. This is a *risk*, for the reason outlined in n.1 to Chapter 1.
77. I owe some of what I have had to say here to conversations with Phil Hutchinson and Tom Greaves, to whom many thanks. Thanks also to Peter Kramer, Gary Francione and Sergio Fava for comments on a draft. And to the audience at the Crewe "Film-Philosophy Festival," October 2011. Thanks finally to my MA "Philosophy of Literature" students, for helping me thoroughly test this material.

4 When Melancholia Is Exactly What Is Called For

Melancholia (and *Solaris*)

> 6.43 If good or bad willing changes the world, it can only change the limits of the world, not the facts; not the things that can be expressed in language. // In brief, the world must thereby become quite another, it must so to speak wax or wane as a whole. // The world of the happy is quite another than that of the unhappy.
>
> 6.431 As in death, too, the world does not change, but ceases.
>
> 6.4311 Death is not an event of life. Death is not lived through. // If by eternity is understood not endless temporal duration but timelessness, then he lives eternally who lives in the present. // Our life is endless in the way that our visual field is without limit.
>
> —Wittgenstein, *Tractatus Logico-Philosophicus*

> If I take death into my life, acknowledge it, and face it squarely, I will free myself from the anxiety of death and the pettiness of life—and only then will I be free to become myself.
>
> —Heidegger, *Being and Time*

The chapters in this book are all, as a Paul de Man might put it, allegories of my viewings of the films that this book places centre-stage.[1] Those viewings are, I hope, "therapeutic" in the best—Wittgensteinian—sense of that word (see *Philosophical Investigations* 133): the films discussed have the capacity, that is, to *free* one from what holds one intellectually/emotionally/politically captive.

In fact, then, we don't need to stick to the word "therapeutic," which for some readers will have the wrong connotations. We can, as I've noted, substitute words such as "freeing" or "liberatory." This is how a freeing film-philosophy, of enlightenment, is arrived at. One is no longer a victim of one's history, one's culture, one's psychology, one's assumptions.

This chapter is an account or allegory of *my* viewing(s) of Lars von Trier's remarkable film, *Melancholia*. It is personal and philosophical, where the personal is at the same time philosophical. Von Trier's film, in turn, is a brilliantly illuminating allegory of (his) depression—and it is also much, much more than that. In expressing my experience of the film and the world (and my experience as a part-time melancholic, which is part of my basis for describing the film as "brilliantly illuminating"), my essay is *inevitably* personal, "person-relative." This is a central feature of therapeutic/liberatory philosophy, practiced most famously by Ludwig Wittgenstein; as the later Gordon Baker explained clearly: such philosophy "responds" to the reader (or viewer).[2] And *vice versa*, in a kind of dialogue or—to use the term that *Melancholia* prefers—a *dance*.

The fundamental reason to think *Melancholia* philosophical is that the film centrally concerns how one can preserve/create meaning in one's life in the face of perhaps the most complete catastrophe imaginable, which is *inter alia* a cultural catastrophe. It also focally concerns, in a micro version of that macro phenomenon, the difference, subtle and narrow yet crucial, between depression on the one hand and grief or deep sadness on the other: the pathological nature of the former versus the intelligibility and called-for-ness of the latter. Thus it might be called a work in the "philosophy of catastrophe": both cultural (political) and individual (psychopathological). Macro and micro. It uses the micro as a route to understand the macro, much like, in other chapters of this book, I argue that films such as *Hiroshima Mon Amour* and the *Lord of the Rings* trilogy—and, in fact, most of the films in this book—do.

There is also, however, a less obvious way to understand how the film is legible as philosophical. This more roundabout route might helpfully start with a strikingly odd plot feature of the film: the entire action of the film takes place within the grounds of a family home, a chateau. This is, in part, because Justine—and in fact, later, both the two main protagonists—apparently *cannot leave* the chateau.[3] Each time Justine attempts to take her horse across the little bridge away from its grounds, she fails. Near the end, the same uncanny failure (which previously affected Justine alone) hits Claire, in her golf-buggy, the last vehicle able to move (albeit with the risible speed and style of a golf-buggy) in their little world. I believe that this uncanny trappedness is a key to the film. What does it mean?

We can think here of *Last Year in Marienbad*, so clearly intertextually telegraphed in the opening images of the sculpted plants outside the "chateau" which have two shadows (while in *Marienbad*, in mirror-image to this, they cast none at all). In *Marienbad* too, it is impossible to escape the chateau and its gardens. One is trapped, on my reading of that marvellous and puzzling film (see Chapter 2), in one's own unreasonably overweening Reason. In one's (in the film's characters') own half-dead hyper-rational mind, which equals to being trapped in psychosis, as understood roughly along the lines envisaged by Louis Sass (in his *Madness and Modernism*).[4] Psychosis is being trapped in and by one's own mind without sufficiently realizing that one is so trapped (because a full such realization would *ipso facto* break the trap). In the case of *Melancholia*, we are dealing primarily, it would seem, with neurosis, or with "affective disorders": being trapped in one's own thoughts and their concomitant moods, and knowing it, and hating it. The trap, in this second (similar and dissimilar) case is simply (in) one's life. The trap is one's mind. The chateau is a lived world. The chateau is your mind. You cannot escape it. (At least, not directly.)

Claire's effort to "run" away are magnificently depicted in their utter futility (because our machines are not going to save us),[5] showing her attempts to escape first by getting into a big strong car—a 4x4—then a golf-buggy, and then just running. But where? The interaction at that point between her and Justine is startlingly reminiscent of the interaction between Deckard and Roy at the climactic moments of the famous chase in *Blade Runner*, as the latter asks the former, as Deckard seeks pointlessly to escape his fate, his being-towards-death: "Where are you going?" Justine uses the exact same words, to Claire. The point, in both cases, is that there *is* nowhere to run to. There is no escape. You cannot run away from your life, nor from your ownmost death, nor from the present moment (somewhat as Leo later puts it, at a pivotal moment in the film: "Dad says there's nothing to do then. Nowhere to hide"). The only "escape" from what Freud called "ordinary unhappiness" / anxiety and, still more so, from melancholia, is (as Buddhist thought has long indicated) equanimity/*acceptance*. To "escape," paradoxically, one has to embrace: to accept what is happening right now, to embrace it, and to embrace others as well. This, as I will discuss below, is what Justine at this point in the film is managing, for the first time, to do, in both a very direct and a symbolically rich way. There is no (other) escape. This existential point, which we'll also examine in Chapter 6 in connection with the architectures and psychologies of *Lord of the Rings*, is yet more starkly literalized in *Melancholia* by our whole planet being about to be wiped out.

The trapped condition of the characters is signalled in the film through a number of stark, aggressive symbols, especially in the exquisitely crafted prelude to the film proper. The most obvious of these is calling the huge blue planet about to crash into Earth "Melancholia": a bald metaphor for Justine's condition. Also striking in this context are the scene of Justine walking through the forest in her wedding dress, so, so, so slowly, held back by the creepers (this scene, we later discover, is a direct representation of Justine's experience, as she attempts to explain it to Claire); and the parallel scene of Claire, seeking desperately to carry her boy Leo "to safety" across the 19th (!) green,[6] but sinking in so deep with each infinitely slow step. These are visual metaphors of/for the mental states from which the sisters are, hopelessly, seeking to flee (and thus inadvertently entrenching—see below).

Other visual metaphors point to a more hopeful resolution. When one thinks back to this "prelude,"[7] from the end of the film, one notices that virtually none of the scenes it contains are present anywhere in the body of the film. For instance, the scene showing the final trio of the film, standing, facing the camera, *separate*, on the lawn at night, dressed up in their wedding gear, but with the two "moons" (our Moon, and Melancholia) behind them (as was not yet the case, during the wedding). The scene almost looks like a publicity still for the film. This is how the three of them would have been, had Melancholia come to hit on the night of the wedding. Apart. Before the journey on which Justine leads herself and them, through rock bottom, to mutuality and an affirmation of life made directly in the face of mortality.

It might still be claimed that, once one notices them and thinks about them, these "visual metaphors" are rather stark, too obvious. There are, however, less obvious ones, which help us understand the real meaning of the film, showing the subtlety lying behind some surface un-subtleties. The deliberately plodding telegraphing of one or two of the film's central metaphors can be seen as the counterpart of a much subtler, sinuously delicate way in which metaphors that are not merely literalized, not straightforwardly paraphrasable, enter repeatedly into the film. Precisely because of the blatancy of some of the basic symbols/metaphors of the film, *these* latter are by contrast easy to miss.[8]

Let us turn for instance to the question of why Part 1 of the film is called "Justine" and Part 2 "Claire"; and to the question of why the wedding scenario in Part 1 is so madly over-the-top. As I suggested earlier, the chateau is a world; yet, more than that, what we are given in Part 1 is *Justine's* world. This world is very like the world that all of us live in, in that it is filled with family conflicts and saturated by the influence of rampant capitalism (in this context, Justine's talking back to Jack at the wedding party, her speaking truth to him, represents real progress). And yet this world is also in many ways unlike the world most of us are familiar with, except for those of us who are personally familiar with a serious amount of melancholia/depression. One can sympathize with why Justine would be so troubled, when one meets the especially crazy, "normal" people in and governing her life, including crucially the people who brought her up; but the situation is more complicated than that, underdetermined by it.

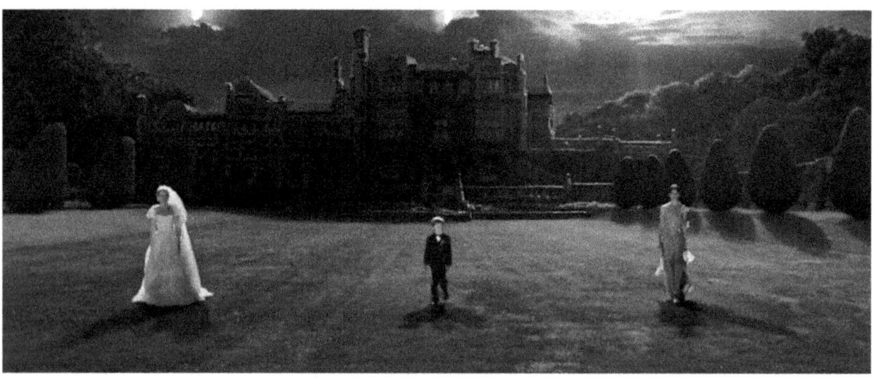

Figure 4.1 Justine, Claire and Leo, in the 'Prelude' to *Melancholia*.

The arc of the journey the film takes one on is closely tied to a complex sequence of one's identifications and dis-identifications with Justine (and then with Claire). This leads to an implicit delicate play with the meaning of "world" in the film, which is the direct counterpart of, or complement to, the deliberate plodding in the Melancholia-as-a-world-perhaps-about-to-smash-into-our-world metaphor. Here is a sketch of the main elements of this sequence as one finds them in the first two hours of the film (abstracting again from person-sensitive issues such as one's experience or otherwise with melancholia):

- From the start of the prelude, Justine is an Other, a haunted figure.
- Then, at the start of Part 1, she seems perhaps just a normal "gal" after all, a normal bride. (Look at her giggling in the car at the failure of the chauffeur to get the limousine to penetrate its way up the chateau's winding road.)
- But we come to see gradually that she *is* haunted. That she has been putting a brave face on things, her smiles largely a (sometimes bravura) performance. It is perhaps understandable that she should behave in such a way, when one starts to appreciate her (largely dreadful) place within her (largely dreadful) family, job, life. (Her depressed mood is first brought on, or at least brought out, by her parents' truly terrible "wedding speeches.")

The film explores the reasons for, *as well as* the unreasoned-ness of, depression: it is not as if Justine's dismal life is enough *reason* to be permanently melancholic, though one can well see how it might (and does) feel that way. On the contrary, we eventually realize with her that even in the valley of the shadow of death there is every reason to feel love and even something along the lines of joy. To escape the confines of one's mind as it has been. The film is an increasingly convincing (as one watches it, as it goes into depth) portrayal of melancholia or "depression." Of how it is based on something—and based on nothing; and of how it can be accepted—and thus overcome. The film undercuts the absurdity—the widespread, ghastly illusion—of the idea that one can be "made" happy by things (especially, by *things*).

We can think here of the opening of the U.S. Declaration of Independence, which famously states: "We hold these truths to be self-evident, that all . . . are created equal, that they are endowed . . . with certain unalienable Rights, that among these are Life, Liberty and *the pursuit* of Happiness" (italics added). It would be ludicrous if it read ". . . life, liberty and the attainment / full realization of happiness." One might well have a right to life or to liberty. One certainly has no *right* to happiness. This is a key point that I see *Melancholia* as enjoying making: the idea that there is either a *right* or a *responsibility* to be happy (or both), a kind of psychologizing of liberal ideals in the context of a consumerist society (the kind of society America has *become*), is itself, von Trier is suggesting, a central *cause* of our unhappiness. The fantasy that we ought to be able to guarantee the avoidance of melancholia *is itself a central cause of melancholia*.[9]

Over and over, even into Part 2 of the film, characters insist that Justine *ought* to be happy; and there is endless talk of Justine being *made* to be happy. Claire's awful remark, "You're lying to us all," captures the double bind perfectly: it is not good enough for Justine to act happy; she has to actually be happy. She "*must*" be authentically happy. John's conversation with her along similar lines is equally extraordinary and telling. The skin-crawling ghastliness of the scene where the bride is supposed to toss her bouquet, with the uncomprehending smiling faces of those staring up at her at this point and braying at her, is a lovely (sic) visual version of this. The point, we eventually understand (and experience?), is that, when one really lets go of the counter-productive effort to project a state that is not one's present state, only then can one start to attain a kind of contentment, a joy in the moment.

- Justine is othered, then, in her depression, and we keep veering back to her, in our recognition of the madness of her (our) world.[10]
- But we gradually come to appreciate that the wedding party is a hyperbole; it is not even *meant* to be realistic. This is most stark in the behaviour of Justine's boss Jack (as grotesque and cruel as anything out of de Sade—and he is the "best man"!) and his minion, Tim. This is a kind of Kafkaesque absurdist extreme of no escape: the profit motive and a rigorously utilitarian attitude to other people will not leave one alone for even one moment, not even at one's wedding. The crazily exaggerated party scenes are rather meant to give us some insight into our market-mad world[11] by touching uncomfortably on what might well be claimed to be its contemporary essence; and at the same time to give us some insight into Justine's world. In a world of depression-retreat, of being locked in one's own ego, and in times of high anxiety, everything can seem extreme: too much trouble, such that one cannot even lift one's leg into the bathtub; or everything can seem like a tremendous threat that cannot be overcome. The wedding party gives us Justine's world: we eventually inhabit that world (as if) from the inside. We realize something about the world of the unhappy; just how deeply it differs from the world of the happy.[12] The only way to come to see something like that is an extreme vicarious experience: such as that of a wedding-from-hell which is really a wedding in hell (hell being not, as one of Sartre's characters said, other people, but rather, contrariwise, the felt *absence*, the unreachability, of other people, even in their presence).
- While Part 1 is Justine's world, Part 2 adds into Justine's world—which now, from a complicated dance of outside and inside, of actuality and

possibility and (possibly) impossibility (*Could* a human be quite as bad as Justine's boss?), as we see Justine in her fuller abjection, we come to know—Claire's world. In Part 1, we probably did not like Claire terribly much. In Part 2, we come to appreciate the terrible difficulty of living with someone like Justine (and with someone like John, her own husband!). We come to appreciate Claire, her patience, her love. We come to know and be touched by herself, her world, including her ordinary unhappiness, the ordinary anxieties of life (Claire, John tells us, "gets anxious so easily"). She is probably closer, for many viewers, to being a natural avatar for oneself.

- In Part 2 we also gradually come to understand how inadequate Claire is to the threat of death. Two worlds may be about to collide—not just hers and Justine's, but also Earth's and Melancholia's. The second part of the film is no more (and no less) "realistic" than the first part. It is a deep engagement with "the reality principle," in the shape of utter vulnerability, death—and its denial. This blue planet, our double, which shows us (from the prelude sequence onwards) the arbitrariness of our placedness and "security" in the universe, and which crashes into us in spite of (or, metaphorically, because of?) our best efforts to master nature, is in this sense no less (but also no more) unrealistic than the wedding party of Part 1. Just as Justine earlier struggled with it, so Claire, with all her caring nature (which for a while, when John takes the coward's route out, leads her to become the main caregiver), cannot ultimately cope with the impending arrival of Melancholia.
- Earlier we pitied Justine, and tried to empathize with her. But our position was no more secure than hers. This is what Claire's arc tells us. Facing death, being-towards-death, is a near-impossible challenge.
- Yet, we want to rise to that challenge. We want in this way not to be Claire. Gradually, in Part 2, there *is* something to fear (which there was not, in Part 1, and yet angst was there, uncanny, massive). Claire majors on (ordinary) anxiety, ordinary unhappiness, rather than depression; but these are not so very far from being two sides of the same coin, two worlds that can be seen more clearly in the reflection of each other's image. In the situation now unfolding, in the "dance of death," without undue attachment to life and to desire, in the dance of Claire and Justine, the depressive sister is now the better off. This is the film's distinctive contribution to investigating the "ecology of depression": in a certain "niche," depression is adaptive (I will return to this point). As Melancholia approaches, melancholia ebbs, or at least becomes non-pathological, and provides an opening to reality, including to empathy. The planet is what occasions the bringing of something to a head: the proper awareness of the preciousness of this timeless moment.
- Thus as Part 2 proceeds further, we avert from Claire and swing towards Justine again. She now becomes the well-adapted one, in this new environment, this new world-with-a-deadline. The change begins with the

key moment at which, at the end of their second horseride together, Claire and Justine see Melancholia for the first time. From this point on, Justine is able to cope with the presence of her depression/melancholia in a way that she was not, previously. And suddenly, her *fear of* melancholia gone, or her melancholia *at* her melancholia gone, she finds melancholia far easier to cope with. The "second arrow," as Buddhists would put it, is no longer present.[13] She is no longer caught up in the pattern of being scared by her condition that so marked her in Part 1 of the film, and that prompted her reachings-out (completely unrequited) to her parents. Once the planet is there, visible—rather than merely being this off-stage thing that Justine is preternaturally aware of, as that which blocks Antares; rather than merely an unseen threat—its sting is gone. For Justine, but not for Claire. Justine suddenly becomes the well-adapted one, and we gradually become impressed by this.

- This process too, however, needs to be interrupted, for Justine is still caught up in an unhealthy state of mind. She sometimes *wants* life to end.[14] She is relieved by the prospect of the world coming to an end: now she is—at last—able to live! Our attraction to her hatred for the Earth / for life is of a kind with our attraction to her very psychopathology.[15]
- This shows that we have not yet found what we were seeking for, a truly authentic (mode of) life. In fact, Justine's new state is far from being freedom from the confines, the iron cage, of the ego; her nihilistic words to Claire may momentarily attract or impress us, but then on reflection they repel us from her again, and appropriately so. The repulsion is accentuated by her brutality towards Abraham, the film's Turin horse (who is an heir to the similarly beautiful and unpredictable horse who one encounters in Part 1 of *Solaris*). It is precisely at this moment in the film, as she realizes perhaps that there is no escape, that, significantly, she (and we) see the effects of Melancholia for the first time.
- We are attracted by Justine's nihilism; but this is a dangerous seduction that tells us something about ourselves, and thankfully we come to see this as we see that she is not a reliable moral "narrator."[16] We were in denial about her, about our attraction to her—which *is* the attraction of melancholia. We needed shaking out of it.
- At this point, we could perhaps start to dialectically synthesize what is needed. We can put together Claire's caring nature, her passion for life to go on and for her child to have a future, with Justine's calm acceptance, her refusal to pretend, her presentness. The sisters could be, together, one person waiting to be born, waiting to be the child, the community, the future. This is where you (the viewer) come in.
- The journey is not yet over, though. In the very final stages of the film, there is a rapid sequence of further shifts, crucially tied into Justine's emergence as a brave and almost heroic, loving, feeling, quasi-maternal figure, when tested to the limit. I shall explore *this* sequence below.

These points, then, sum up the real, subtle meaning of the apparently overly literal metaphor of Melancholia as another world, another blue world coming to meet ours. In the meeting, we (the film's necessary other: its audience) find ourselves: we find what kind of world we have, triangulating our world with Justine's and Claire's. It is us, sitting with them in the "magic cave" at the end. We could imagine a Part 3 to the film, about another character in it, and then a Part 4, and so on and on until every human being in the world had had their world added in.[17] But we do not get to experience this, because there is not enough time in life to get to know everyone in the world, let alone in a movie. Our lives dance an arc that ends with death; and sometimes this death comes much sooner than we had hoped/expected. This is what we have to live in authentic relation to. There is no Part 3 to the film, because, suddenly, the(ir) world ends. As yours and mine of course will, much as we are too often in denial about the fact. (And as our world as a whole will, unless we learn the kind of lesson essayed in this film / this book.)

It is difficult to understand another's world, but it can be done. Sometimes, in order to do it, one has to take a circuitous, "indirect" route. In very difficult cases,[18] one may have to take a deeply circuitous route that may even take one on a journey through nonsense,[19] a journey through trying imaginatively to inhabit positions that are not even inhabitable. As Rush Rhees put it: language makes sense only if living makes sense. So when living itself starts to deconstruct, when the way we live starts asymptotically to approach the absurdities set out in *Melancholia*, in *Marienbad*, in *Never Let Me Go* and *The Road*, in Peter Greenaway's films etc., then we shouldn't be surprised for language and art to veer necessarily into nonsense.

This—arriving at sense through banging one's head against the *limits* of sense—is a central method of Wittgenstein's in philosophy,[20] and it is also the method of some fine "philosophical" films such as *Melancholia*. A key case of something in this way difficult to understand, despite the fact that it is present to some extent in everyone (especially around the question of death; cf. Chapter 3, above), is denial. This film helps us to understand and work through our own tendencies to denial, and through those of others. In *Melancholia* we find those tendencies and temptations expressed, and at the same time we are offered a way through and beyond them, back to life. In short, *Melancholia* exposes those temptations (on film), and it does so in such a way that we may be better placed to enjoy life while it lasts, and to be clearer about how precious and glorious it truly is—and thus about how we ought to strive to make the human adventure and the existence of our non-human kin last longer than we are currently threatening to let it do.

The scenario in the film is a neat inversion of the situation *vis-à-vis* climate change denial.[21] In *Melancholia*, it is the loons and conspiracy theorists who are right, and the scientists who err.[22] In both the film *and* the real world, it is the "pessimists" (i.e. the realists)[23] who need listening to, if there is going to be a wake-up call that leads us to do something adequate about our predicament. Back in the real world, back outside the movie theatre, there is in effect another planet smashing into us; just very, very slowly. What anthropogenic climate change threatens is the (gradual) equivalent of the comet that extinguished the dinosaurs. We have to perceive this, our clear and present long emergency, as an emergency even though it does not feel like one.[24] Part of the achievement of *Melancholia* is to depict the looming destruction of human life on Earth—the risk that life on Earth is, as Justine memorably puts it, "not for long"—as an emergency (as urgent; indeed, as rapid. Recall the rapid onset and progress of the freak weather at the end of the film, reminiscent of the climate-induced weather-chaos that we are experiencing more and more of).

And yet . . . also not as an emergency. The feel of so much of the film and of the discourse that occupies it is far from any emergency talk.[25] From the partying of the first half of the film to the distantiated feel of some of the second half, *Melancholia* reflects and thereby problematizes inaction in the face of impending disaster. As part of its call to authentic affirmation of life in the present moment, we should hear *Melancholia* as calling us to fight to prevent the rich elite destroying the Earth as a liveable planet, by saving its atmosphere, keeping it life-giving.[26] *Melancholia* is a warning about how we are inclined not to treat even the worst thing that can happen as an emergency.

The way that Justine pathologically experiences the world—her world, this micro world that gets shown us—*shows* us essentially the actual, pathological nature of the macro world we have made. The terrible threat she semi-constantly imagines/creates/experiences psychologically IS real, ecologically.

"What's the worst thing that can happen?" This ordinary, helpful question, very necessary for us when we are seeking to head off uncertain risks that may destroy us, becomes less helpful, precisely through seeming to offer deep help, to the person inclined to depression / severe anxiety. The reason is that such depression/anxiety *is* precisely the imagining, over and over again, of what the worst thing is that can happen.[27] It is a would-be self-protective race to the bottom. One seeks to immunize oneself against the future by giving up hope for anything good; one seeks to protect oneself against other people by imagining that they think the worst of one; one seeks to protect oneself against hope for oneself by imagining oneself hopeless/useless/evil. These stratagems are extraordinarily seductive;[28] but they are also, in the end, disastrously self-defeating. One cannot actually become safe by retreating away from others / from hope.[29] One seeks to immunize oneself against disappointment by pre-emptively disappointing oneself (and others); but this only ups the ante, and takes one on a journey deeper into the morbid life-world of melancholia. The desire to be immunized against hope, the desire for disaster to absolve us of responsibility and to protect us pre-emptively from disappointment, is the very same desire in politics as it is in psychopathology,[30] in macrocosm as in microcosm, in Part 2 of the film as in Part 1. It is the desire that *Melancholia* explores (and, ultimately, explodes).

"What's the worst thing that can happen?" Followed perhaps by, "It's not the end of the world!" But, what if it *is* the end of the world? *Melancholia* splendidly literalizes this central trope of depression/anxiety, which is also a very necessary trope of ecology. Yet another way then to see clearly how this film is a work of philosophy, not some post-Modern twisted romcom, nor a pseudo *cinéma vérité*, is to look at the peculiar way that, throughout the film, steadily, Justine "knows" things, symbolized by the 678 beans in the jar at the wedding. What is it that Justine knows? I would tentatively suggest as central to what she knows, which most of us either do not know, or know and yet desperately try to go into denial about, is that the society of the super-rich that she is seemingly trapped in is sick, that the way in which neo-liberalism in general and profit-oriented business in particular treats individuals is unacceptable and catastrophic socially and ecologically, that depression (or at least deep and persistent sadness, grief) could even be a rational response to such treatment. She knows that the complacent way in which we tend to imagine that our ways of treating each other and our environment will not bring life on Earth (or at least human civilization) to a state of permanent decline or indeed terminal destruction needs puncturing. Like Elle early on in *HMA* at the moment when she prophesies that we'll use nuclear weapons again (unless, possibly, we manage to remember, to see), Justine is a needful kind of Cassandra for our times:[31] seeing reality with the mixture of cold realistic hard-headedness and desperation that can be characteristic of depression, and warning us urgently about it. Warning us about just how bad things are, and about how bad as a result they could become. Seeking slumber's end.

Claire says, diagnostically, to Justine, in a key scene in Part 2: "It's easy for you, isn't it? Just imagine the worst thing that can happen." Claire now sees the attraction of Justine's world. The attraction, the would-be safety, of imagining the worst thing that can happen: death, "for example." The death of everything, in fact, geocide/ecocide.[32] This explains to a considerable degree the attraction of disaster movies, of apocalypse movies:[33] we imagine that if we can cope with these experiences fictively, then we can cope with them more easily in real life. Thus we half-want the world to be destroyed (in this film). We want it to go up in a shriek, and us with it. (And we half-half-want the worst thing that can happen to have already happened in the real world too. For then, we can be absolved of responsibility, of hope; we are given an indefinite *reprieve*[34] from having to try to act, and can simply spectate.)

Eventually, thankfully, we pull back from this. We pull back from it at the very end, in terror and horrified awe, with Melancholia bearing down remorselessly upon them/us. Only, now, it is too late—and this is a crucial part of the film's "therapy."[35] At the final moments, at the death, one does not want ge(n)ocide any more, not even fictively. One wants life. And this is why what Justine knows has an opposite character from that cynical knowingness that characterizes so much of what passes for culture today, including that on "the Left" and amongst the intelligentsia. To face reality and, rather than suffering / indulging in nausea, to want life, to be *in* the love, to be truly present in it and in our predicament: this is the challenge that is far too often given up on, by those who look to immunize themselves against the grave risk of disappointment. The culture of complacency that animates the scientists who refuse to sound the alarm about the state we have brought our planet to, the outright climate deniers who seek to woo us back to sleep with the siren call of a pretence that all is well if only the messengers would shut up, the culture that animates too the doom-mongers who create a sense of such hopelessness about the prospect of doing anything that they *make* the very outcome that they are claiming is inevitable. . . . It is to this kind of culture that Justine offers an antidote.

What could be more depressing than the end of the world, especially by our own hand? Given this situation, depression (or rather, at least, deep sadness, persistent grief) can indeed seem the appropriate and *rational* response. I am thinking here of the "ecopsychological" vision of Theodore Roszak and Mary-Jayne Rust among others.[36] The idea that one's (negative) mental state may well be a reflection of the physical/biological state of the world that one is sensing here coheres with the sense I have of von Trier's film. (This is an idea that takes us decisively beyond the fantasy of individualism that characterizes most psychiatry and psychotherapy.) What is alarming, then, is that so many of us are *not* at all depressed (picture for instance the non-depressed people who predominate in Part 1 of the film: so much the worse for them, for us). In *Melancholia*, the depressed one is not made depressed by the end of the world. On the contrary. Claire is the normal one, who gets thrown into anxiety by it; thus by the film's end it is she who is trapped in the grounds, trying but unable to leave.[37]

By the time enough of us get anxious/depressed about what we are committing the ecosphere to, it may well be too late. People mostly get depressed individualistically. Collectively, we watch the build-up towards destruction of our planetary home as a spectacle, alarmingly un-depressed. That is how it is, implicitly, in *Melancholia* Part 1 (It only changes during Part 2). It is almost as if we are *willing* the devastation of our home.[38]

Thankfully, Justine too pulls back from this wish for death, at the moment of death's final approach. The wonderful thing that happens towards the end of the film is that, under the most extreme pressure, with the worst thing that can happen now utterly inevitable, and with hope gone, at last Justine manages to emerge into living in the present, with others. The embryo of this re-emergence[39] can be seen by the time of her authentic and clear rejection of Claire's unconvincing plan (a plan that echoes the dead rituals of the wedding, in Part 1 of the film, a plan that only looks good to someone who thinks that one can be *made* to be happy[40] or can make oneself happy; e.g.

by booze, by polite convention) for them to drink wine together on the terrace as Melancholia hits. But the emergence of Justine's new ability is truly born only with her embracing of the boy Leo, her decision to care for him in his fear. The crucial moment here—the film's greatest turning point—is her crying as she hugs him. She breaks—as one might put it, reversing Bob Dylan—like a woman, like a heroine, this time. The whole weight of the film is in this scene and the next one, in the final seven minutes of its running time. Justine breaks non-selfishly, for the first time in the film. Leo cannot literally see this (but we can): she is feeling *for him*, grieving for someone other than herself, for the pain and the shortness of life *of the child*; she has managed for the first time truly to escape the terrible confines of her own mind, the iron grey cage in which melancholia can hold one; she is back in empathy. Whereas earlier Justine was mostly unable to function as "Aunt Steelbreaker," lacking the extraordinary human ability to make things OK by being present and confident (that is: with faith, of a broadly Kierkegaardian/Jamesian kind), and by art (metaphor), now she can manifest love and care. She can do something that seemed impossible, like breaking steel with one's bare hands. Her finest hour is her last: she gives Leo a blatant metaphor; she tells him a story, and they will then build visuals to go with it in the shape of the "magic cave." What we see in this scene is perhaps how adults should face authentically what is happening, while younger children should be protected. For this reason Justine builds the flimsiest structure imaginable, to symbolize the truth: that *if one can only live in the present, one is perfectly, absolutely safe*. She makes the "magic cave" with Leo, then beckons first him, and then Claire, inside, and then joins them, having "completed" the cave. Here, crucially, we see Justine supporting Claire, in a visual mirroring of the scenes where Claire supports Justine's movements, in the opening portions of Part 2. This role reversal is an iconic image of the teaching that the film offers.

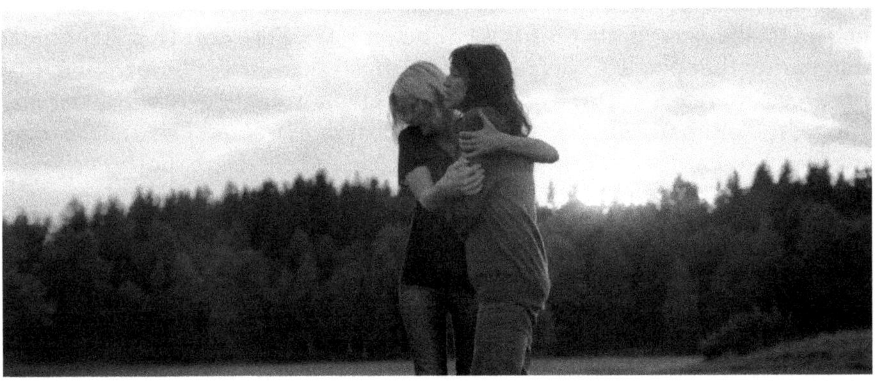

Figure 4.2 Justine's turn to support Claire, as Melancholia closes in.

In this very moving final scene, we watch them gradually holding hands, as the Wagner "prelude" music swells. The climactic moment is the *wonderful* courageous smile that Justine gives Claire (at 2.02.25). The smile that tells that she is having, at last, what might be called a wonderful life, even amidst real and psychical horror. It is a smile of love, of genuine connection, genuine being-with, at last. She looks authentically into another's face, for the first time able to do so while offering something authentic that is not (only) sad.

Justine's smile at this pivotal moment is—with the possible exception of her smile earlier at Leo—the first real smile she has given anyone in the whole movie. This authentic, fragile smile comes from the depth of her whole being, in the awareness of life and its fragile preciousness and beauty, rather than being (as her smiles earlier in the film were, at best) an isolated moment of relief, or (more often) a mere show. This, by contrast, is the smile that reconnects and, in a sense, reassures. For now, at last, she is paying *attention.*[41] Being-with-others. Looking from Leo to Claire, being supportive and present with them, feeling for them. Her lower lip juddering/quivering just slightly, and then the loving, surprisingly strong, attentive smile. She, and thus potentially those with her, including us, are now very close to what Wittgenstein is referring to in his "Lecture on Ethics," when he speaks of "the experience of feeling absolutely safe. I mean the state of mind in which one is inclined to say 'I am safe, nothing can injure me whatever happens.'"[42] We have come very far at this point from the fake safety of withdrawal, perhaps of paranoia, of hopelessness, certainly of pre-emptive disappointment, that marked Part 1 and the earlier portion of Part 2 of the film.

Melancholia is about one's desire to escape from one's mental state (a desire that forms the central mechanism of depression and anxiety), the desire for a reprieve from one's responsibility to face up to one's situation and act accordingly, and it is about the only genuine security there is: being present, no matter what. What is critical about Justine's re-emergence is that it gives the lie to her brutal negativity a little while earlier. At the moment of death, she overcomes the vast, horrific temptation to think that the Earth and all its life are evil, that it would be better if we were not. That is the great temptation that plagues Nietzsche's Zarathustra: nausea at humankind. Justine, instead, reaches out, lives and loves, giving those she is with, Leo and Claire, the ultimate gift: her presence, at last, *in extremis*. This *is* authenticity; being-towards-death. It means being able to face death, not to be taken by it, as Claire in the end is. Claire, with support, does her best, but at the very last she cannot continue; she withdraws, into herself, in highest anxiety, covering her ears. Justine and the boy, instead, keep going. This is their finest hour. The rising climax, at the very end of the film, is staggeringly moving. Justine has completed the journey, the arc, from the relative inauthenticity of a truly desperate struggle, to authenticity and being-with-others. Her ability to be attuned to Claire's and Leo's being, in the final minutes of the film, her being not just a rock but a fellow being for Claire, giving her true attention moment after moment, show us a model of mindfulness and care for anyone to follow.

This life-affirming finale is however tinged by sadness, not only because of the impending end, but also because of the awareness, as Justine's poor unconsummated husband Michael puts it earlier in the film, that "It could have been very different." While we finally take in the possibility of living in the present and overcoming the vast reasons to withdraw / to give up hope / to live in fear / to be a slave, Justine and Claire might feel a twinge of regret that such living in the present could not be achieved beforehand. If only one could live one's whole life like that. This Wittgensteinian/Heideggerian/ Buddhist ambition is what the film offers us as a possibility.[43] Death may come at any moment. Let us live in that moment, i.e. *this* moment, authentically, and smiling a real smile if we can.[44] The viewer is invited, I would say, to leave the auditorium ready to smile such a smile, and to live as Justine lived at the end. Despite initial appearances, there is no pseudo-Wagnerian love-death as such in *Melancholia*. In fact, at the end, Justine is perhaps even a little in love with life, as may be signalled by the fact that, as the big blue planet bears down on them, in one last great metaphor, Justine calmly *turns her back* on Melancholia. She loves her family, she loves human beings, at last, at (the time of) the death (of all things). Thus she steps out of the victim role.

Nevertheless, one could worry that the film's dalliance with apocalypticism is dangerous. This worry is dispelled if we view the film, rather, as a *commentary* on apocalypticism,[45] roughly (as I argued earlier this chapter) as it is a kind of commentary on metaphor, a critical examination of it. In doing so, the film enters a therapeutic and critical relation with our desire for the world to end, for the worst possible thing that can happen to happen, our desire, perhaps even, to "cleanse."[46] This desire is not entirely to be dismissed: one wants an end to suffering, a quick release from the slow journey of news and torments that may be the human race's downgoing, perhaps even a cleansing from the Earth of excessive numbers of consumptive beings that are in effect consuming/destroying its life-bearing powers (the process turbo-charged by the very advertising industry that the Dunst and Skarsgård characters, Justine and Jack, work in). Yet one also comes to see that even if such destruction were rapid, it would still be utterly terrible; and one comes to understand (as Justine does) just how precious and wonderful life is, and how bloodless it is to be ready to give up the human adventure.[47]

Still, a worry remains: Is it not wrong that the end of the world should be depicted as, arguably, beautiful? It is indeed utterly awe-inspiring, *magnificent* in the true sense of this overused word, when we see Melancholia bearing down on us in all its hugeness, in those final moments. But the right way to see what happens there also includes the sad but utterly understandable inability of Claire to stay with the trio in the final moments, distracted by the real fear evoked by that sudden, rapidly growing crescendo of noise—while the viewer in the cinema truly feels like she is present at an earthquake, like the disaster is . . . *real*—and it includes the way in which, after the shock waves, and before the enveloping blackness, comes a fire, burning up our loved and loving trio. *This* is not beautiful, it is just terrible. At the same time, the extreme beauty of much of the film *is* the beauty of our life; the wonderful life that finally Justine allows; the "slow-motion" (let us think again of the prelude) moment-after-moment life that we viewers still have the luxury of. The life that our decadent luxuries threaten to undermine the continuation of.

The very end of the film is perhaps sublime, but not beautiful.

After that end, what? Blackness. One hopes for a Part 3. Or at least an Epilogue, like in von Trier's *Breaking the Waves*. But, rather, as in the final lines of von Trier's first great film, *Zentropa*: "You want to wake up. . . . But it is not possible." There is only, after a pause, silent simple credits, still on black. This is the "final" instalment of the therapy, because now one really regrets what desire one had for the world to end. Your wish has been fulfilled, and it is not what you had hoped for. You wish that the world could go on. You wish that their lives could go on, that there could be a Part 3, and 4, and on and on. . . . But it is not possible.

Only, of course, it is. For here you are! Still alive. In a place of voluntary "retreat," but waking up slowly to the world again. In this sense, *Melancholia* is in the end a film "about" the experience of watching a film like this: it is self-reflexive, as perhaps any major "therapeutic" or liberatory work must be.[48] Then the lights come on, and awareness grows that there is a world here / out there. How wonderful, that you can stand up, breathe, talk with or touch your friends with whom you came to watch this film, go out into the wide world. The film delivers you back into life, with an enhanced capacity to live and feel, to be. Perhaps you will now take the chance not to live in ordinary unhappiness, nor in the fated but overcomeable land of melancholia, but will *savour* life. The way, finally, Justine could and did. The world is open to us always doing so, if only we can rise to the opportunity, and on the real blue planet: Earth, not Planet Melancholia. The real Part 3 begins:[49] *your* world, including a world to save. Wake up: time to live.

***

Melancholia is a richly intertextual film. I mentioned above its evident relationship to *Last Year in Marienbad*, and a relationship that could plausibly be drawn also to *Blade Runner*. But the most fruitful connection of all is to be drawn with Tarkovsky's *Solaris*, *Melancholia* appearing as von Trier's deliberate engagement with the legacy of Tarkovsky, and even as a bold effort to do the seemingly impossible: to improve on Tarkovsky explicitly philosophical work of 'science-fictional' genius.

Melancholia indeed demands to be read alongside *Solaris*, firstly through its reworking of a symbolic engagement with Bruegel's *Hunters in the Snow*. It also demands it, of course, because of its beautiful venturing out into deep space and to another planet that allegorizes emotional turmoil and mental ill-health. And it demands it, lastly, I think, in terms of the need to think side-by-side about how the two films end.

First, and perhaps most crucially, then, there is the Bruegel, a painting explicitly central to part of the plot in *Solaris*. The strange scene in *Melancholia* focusing intently on this picture, a scene presaged by a remarkable reworking of the painting itself in the course of the film's prelude (see below), is yet another indication of how it would be a grave mistake to try to take *Melancholia* as a hyper-realist piece in the way its handheld camera techniques might seem to demand. In the scene in question, Justine, depressed by the horrendous nature of her interactions with her family and boss, by the rituals she is caught up in, and, in a classic double bind, being told that she *must* enjoy *and* must enjoy *authentically*, takes a moment by herself in the "library." Then she notices the way that all the books, on modern art, are open on pages of high abstraction. One of the most fundamental things that she knows is that something is profoundly wrong with this de-realized, devitalized way of seeing the world,[50] which is a way of seeing/being in the world that is directly familiar to those who have experienced severe versions of depression. Suddenly angry, she rapidly replaces these abstractions with moving paintings such as the Bruegel, and the famous pre-Raphaelite image of Ophelia melancholically floating off to her death. These beautiful (but tough) scenes are a kind of representation of the nature of life, including its most challenging aspects; as opposed to its mechanicized absence, evoked by abstract modern art.

But what of the specifics of *Hunters in the Snow*? Why this painting in particular? Perhaps the most obvious point to make is that the painting features "*two worlds*."[51] The micro world in the foreground, of the failed hunters, tired, disconsolate, bringing with them nothing to eat. And the macro world in the background, of partying and festivity. The latter could be thought of as Earth, as pictured in *Melancholia* (and perhaps *Solaris*), especially in Part 1 of both films (Yes, *Melancholia* just like *Solaris* is divided explicitly into Parts 1 & 2). The former as expressing an unwanted psychological reality—*and* the eco-physical reality of the Earth, that the partiers deny.[52] The painting could also be seen, in the context of the film, as implicating the two human worlds within our world: the partying world of the rich, of the chateau, and the labouring world of the poor, of the village that (in *Melancholia*) we never directly see.

Amitav Ghosh writes, of the unhomely yet all-too-real nature of manifestations (in our now no-longer-natural weather etc.) of human-triggered dangerous climate change, that "It is almost as if the mind-altering planet that Stanislaw Lem imagined in *Solaris* were our own familiar Earth: what could be more uncanny than this?"[53] *Melancholia* alters the mind too. So does human-triggered climate change itself. Or at least, it should. As yet, it has not done so enough. Which is why we need Ghosh; and *Melancholia*, *The Road*, *Apocalypto*, *Avatar* and much more.

We grieve, on (above) Solaris, in *Solaris*, with Kelvin (grieving for his lost love, and his lost past, his lost chance perhaps to do the right thing, awfully pseudo-represenced by Solaris). We grieve, at the coming of Melancholia, in *Melancholia*, with Justine, for the microcosmic loss (Leo, and his soon-to-be-ended future) and the macrocosmic loss (Earth itself). Such grief perhaps *enables*—in something like the way that we saw identification with and working through of Elle's grief enabling, in Chapter 2, a facing of the nuclear age and a determination not to let nuclear annihilation happen (again)—a better chance of appropriate grief at the sixth mass extinction that humanity has unleashed, and at the prospect of climatic self-annihilation. Thus setting us up better, perhaps, to work to prevent this. (And here it is an intriguing coincidence that *Solaris* offers us an image of a "flooded" world, inevitably bringing to mind the rising sea levels that are a long-term disastrous consequence of anthropogenic global overheat. And that *Solaris* features a striking polarity between the extraordinary beauty of water and nature on the one hand and the lack of beauty of the techno-industrial urbanism threatening it on the other, most visibly in the sequence of Burton's automatic car driving itself through a landscape of urban speed and alienation. Consider here Kelvin's remark to Burton, at 28.10: "Man is the one who renders science moral or immoral. Remember Hiroshima.")

But to go further in bringing out the *Solaris-Melancholia* nexus, one must of course *look* at the *way* that *Hunters in the Snow* features in *Tarkovsky's* film. Especially, literally in great detail in the early part of the majestic "levitation" scene in the library aboard the spacecraft, a scene in which the protagonists are beautifully but also perhaps painfully free of gravitational attraction.[54] 'Hari' (the protagonist's reconstructed/ghost suicide lover) contemplates the painting in detail, in a lengthy point-of-view shot. What strikes one as she does so is just what *should* strike viewers of this painting: its dance with death. The crows in the trees (which we hear crowing; Tarkovsky brings the painting "*alive,*" as von Trier does too in a different way). The fact that the hunters coming home have brought nothing to eat (except one meagre fox, an animal which is typically regarded as inedible). In this way the painting meshes with 'Hari' in particular and perhaps with *Solaris* in general. But the most striking thing of all for me, in the scene, is what also struck me when I went to see the painting in the flesh recently in the Kunsthistorisches Museum in Vienna: that the first thing she homes in on is something very striking, because very unusual, in such a pre-Modern painting: the one dog among the hunting party that looks out of the canvas *at us*.[55] Breaking the fourth wall, looking at us dolefully (even deathfully?) in a way reminiscent of the opening of *Melancholia*'s overture. For that overture opens,[56] to the opening of Wagner's *Tristan and Isolde*,[57] with the most extraordinary pale, deathful shot of Justine. Very slowly opening her eyes—to gaze directly at us. Then, dead birds start to fall from the sky behind her. Shortly afterward, we see *Hunters in the Snow*—and again, in von Trier's remarkable image, dead birds or other, nameless, things then fall from the sky *in* that painting.

The dog who gazes out at us from *Hunters in the Snow*, and whose gaze 'Hari' so firmly first lights on, is superbly expressive, perhaps more than any real dog. In its gaze, as much as a non-human animal could possibly convey, is something like depression or even despair.

Both these films can be considered analyses of despair. But *Melancholia* treats of neurosis (i.e. "melancholia," mainly). And of how that neurosis uncannily anticipates a kind of lived psychosis that on the macro level grips our world. *Solaris* (and *Last Year in Marienbad*) treat primarily of traumatic grief—and of psychosis.

This is also how *Melancholia* and *Solaris* pull dramatically apart at the end.

The most striking feature of the end of *Solaris* is that Kelvin's coming-home is not genuine: Kelvin is lost in fantasy,[58] in effect stuck in/on/near Solaris: on an island of fantasy (these are based, Snaut tells us, in "islands of memory" in us accessed by Solaris's ocean). An island in its ocean of fantasy. In this way, it is clear that *Solaris* closes very distantly from (for instance) *Gravity* or *2001*.[59] For the odyssey in *Solaris* is completed *only in fantasy*. There is no genuine home-coming.

But it is also therefore, in the end, very distant from *Melancholia*. It is a kind of *opposite* of it. For much of the film, one helpfully might think, as I've said, of *Melancholia* playing neurosis to *Solaris*'s (and *Last Year in Marienbad*'s) psychosis. So, whereas Kelvin—like *Don Quixote* from whom he quotes towards the end of the film, and like the protagonists of *Marienbad*—gets sunk (finally) into a state of ghostliness, alone forever with phantasms and projections, Justine emerges triumphantly from the ghastly state she was in at the end of Part 1 or the start of Part 2. She emerges into reality, and also into being closely *present with-others*.

(The situation then is a little like that which emerged at the end of Chapter 3; that, while *The Road* seems so much grimmer a scenario than *Never Let Me Go*, nevertheless the former has a happier ending. Similarly, it is hard to get much grimmer than the literal end of the world, the destruction of the Earth. But, while *Melancholia* ends with a kind of coming to terms with melancholia which is therefore a kind of triumph over it—for, if one can just be present, then nothing is really completely bad; it is mainly our resistance to what is[60] that makes it bad—*Solaris* ends as desperately sadly, if not more so, as Kelvin's entire time with 'Hari' on Solaris has been. The story of *Solaris* is perhaps the one that really calls for melancholy, in the end.)

Thus there is a wonderful link between these two planets—Solaris and Melancholia—that are really states of mind, and that force us to feel the depth of our "homedness" on Earth. As discussed by Paul Johnston,[61] *Solaris* (like *Melancholia*) challenges our quasi-solipsistic self-preoccupation and anthropocentrism. But *Solaris* ends in pessimism on this front—*unlike* both *Gravity* (see Chapter 5, immediately to follow) and *Melancholia*. *Melancholia* triumphantly turns around the process of *Solaris*. *Hunters in the Snow,* which *Solaris* here takes it cue from, concerns a kind of melancholic dance with or even possibly a giving in to death, or perhaps a delusional denial of it. *Melancholia* offers us in the end a means to "transcend" death in the very face of its reality and imminency, whether on the microcosmic or macrocosmic (or indeed, simply cosmic) scale.[62]

Notes

1. See Paul de Man, Allegories of Reading (New Haven and London: Yale University Press, 1979). To call what I offer an *allegory* of my reading of film means that what follows is, among other things, a *personal*, affective account. At the same time, I aim to bring out a/the kind of experience which *anyone* could have who viewed the film in question with openness and some understanding. That is because, as I aim to exemplify here especially (but also throughout the book), it is possible to write a very personal response to a film while simultaneously making objective value judgements about it and other films. It is clear, at the same time, that the experience will in reality vary to some extent, in part dependent upon the "subject-position" which one has (e.g. one's degree of personal experience of melancholia; perhaps one's gender or class), in part dependent upon one's spiritual or existential starting point, and so on. This is a feature of Wittgensteinian philosophy: the resolution of the apparent conflict between the personal nature of the account and the "objective" claims I make for it is via a proper understanding of how I, following Wittgenstein, take the nature of *philosophy* to be, where the term "philosophical" doesn't come down on the side of objectivity as against the personal. Rather, it bridges or overcomes the alleged gap between the two. This kind of personal response/involvement/continuation is demanded by films like *Melancholia*, as I shall describe, where the viewer is joined with others in an authentic-making "dance," somewhat as (as we shall see) Justine painfully becomes authentic. (Thanks to Peter Kramer and Sergio Fava for help working on this note.)
2. See especially his Wittgenstein's Method: Neglected Aspects (Oxford: Wiley-Blackwell, 2006).
3. This is a famous motif from Buñuel's *The Exterminating Angel*; but also, just before that, in Resnais's *LYiM*: see my discussion thereof.
4. Harvard: Harvard University Press, 1994. See especially Sass's reading therein of The Invention of Morel, a text which helped inspire *Last Year in Marienbad*.
5. I mean, from (say) climate-nemesis. See e.g. Josh Gabbatiss, "Future technology 'cannot rescue' mankind from climate change", www.independent.co.uk/environment/future-technology-cannot-rescue-mankind-climate-change-global-warming-a8187806.html. And cf. n.15, below.
6. As Scott Wilson notes, the film curiously makes more than one reference to a 19th hole, while having Keifer Sutherland claim that there are (as one would expect) 18 in his course. Wilson comments that "The so-called 19th hole exists, but it is not part of the golf course; it is the place where you eat and drink after your round. It is one of the many indications that the *mise en scene* of *Melancholia* is not simply a luxury hotel and golf resort but a more subterranean, phantasmatic domain . . ." ("*Entre*, Apocalypse: *Melancholia* and the eschatological banquet", https://mmmouth.wordpress.com/eating-the-earth-eating-the-world-melancholia-and-the-eschatological-banquet/).
7. The character of the images is somewhat reminiscent of the seven "postcards" which preface each of the parts of *Breaking the Waves*. (Justine, furthermore, is clearly worthy of comparison to the female protagonists of each of von Trier's "Golden Heart" trilogy of films.)
8. And *have* been missed by almost all reviewers of the film.
9. As we shall see below, the arc of the film depicts eventually, at the very end, a *chance* for well-being, in the very acceptance of mortality and morbidity, including mental ill-health, including personal or even global catastrophe. But to attain

this, one has to recognize that anti-melancholic stratagems (e.g. rumination) are frequently the very disease of which they take themselves to be the cure.

10. The brutal way she sometimes behaves as Part 1 unfolds is clearly at least partly forgivable against the background of that world, a world of forced appearances of happiness (see the point above), of horrendous parents, of unimaginably bad employers—of uncaring capitalism and inequality rampant. The way her family (her mother in particular) "do her (head) in" leads her to "express herself" in ways that are not productive—as the ways she expresses herself towards the end of the film, instead, *are*—of any kind of accommodation with life. Before the end, the film also shows intelligently how difficult it can be to be around someone with depression.
11. Somewhat as I argued, in Chapter 3, that both *Never Let Me Go* and *The Road* are deliberate exercises in conceptual exaggeration: rupturing the bounds of what makes sense; though exaggeratedness/implausibility is not the same as—is categorically "less" than—impossibility.
12. Cf. Wittgenstein, Tractatus Logico-Philosophicus, 6.43 again: "If good or bad willing changes the world, it can only change the limits of the world, not the facts; not the things that can be expressed in language. // In brief, the world must thereby become quite another, it must so to speak wax or wane as a whole. // The world of the happy is quite another than that of the unhappy."
13. According to Buddhism, whenever we suffer misfortune, two arrows fly our way. The first arrow is the event itself. The second arrow is the suffering. While the first arrow cannot be avoided, the second one can, being within our control, how we react to the event.
14. This may be the best explanation of the otherwise-weak "soft-porn" scene in which Justine bathes naked in the night-light, seemingly, as it were, inviting Melancholia to come and destroy the Earth. The point about this scene, and about this temptation, this attraction—the thought that she *brings on* Melancholia (just as one actually bears some responsibility for one's depression, for depression is attractive and one can give all too easily into temptation)—is, I think, this: that this is a temptation and attraction *that the film itself offers the viewer*, and perhaps brings out in the viewer. This is one of the film's last temptations, in the dizzying emotional-thoughtful journey that one goes on as the end becomes nigh. The desire to have Melancholia crash into Earth is a desire that the film brings to consciousness in the viewer. But, as I explain shortly: this desire too needs to be—and is—overcome.
15. This connects with Arendt's argument as cited at the opening of Chapter 5, below. One worrying sign that we are likely to allow civilization to collapse and perhaps even life on Earth (cf. n.45, below) to cease, as a result of the slow catastrophe of anthropogenic global overheat, is how open we are to ludicrous technophilic fantasies of exiting Earth for a new home. (For why these are ludicrous, see John Michael Greer's "The terror of deep time", https://worldnewstrust.com/the-terror-of-deep-time-john-michael-greer). I read in these fantasies a not-very-well-hidden hatred of the Earth, of nature—and of the feminine/female.
16. There is a parallel here with the (it turns out) unreliable "chorus" of the native onlooker in Kristian Levring's *The King Is Alive*, one of the greatest of the Dogme 95 films. The Dogme 95 films invariably, much like most of von Trier's oeuvre, share with *Melancholia* a profound interest in altered mental states, deep psychical pain, psychopathology. *The King Is Alive* plays cleverly and therapeutically with one's "right-on" prejudice, one's desire uncritically to believe the exotic indigenous "narratorial" voice, which literally looks down on the petty activities of the increasingly desperate Westerners as they struggle to survive; gradually, as their "journey" takes them to a new reconciled mutuality, through the vehicle of their efforts to learn and play Shakespeare's *King Lear* in the midst of their crisis, they outflank this narratorial chorus of nihilistic words,

and the great words that they speak come to have a great deeper meaning and role, binding them together, healing, acting as the best kind of group therapy. In the case of *Melancholia*, it is Justine herself who eventually overcomes her nihilistic words, as we shall see.

17. Compare my suggestion for how to understand *Hiroshima, Mon Amour* as helping us to start to be able truly to understand the scale of Hiroshima, in Chapter 2, via the thought that the macrocosm is the ensemble of many microcosms. In this connection, these two films are very close.
18. As in the case of Peter Winch with the Azande ("Understanding a Primitive Society," *American Philosophical Quarterly* 1:4 (1964), 307–24), or of Thomas Kuhn with Aristotle ("What are scientific revolutions?," in Krüger et al. (eds.), The Probabilistic Revolution [Cambridge: Cambridge University Press, 1987]), or again of Louis Sass with "the schizophrenic mind" (Madness and Modernism: Insanity in the Light of Modern Art, Literature and Thought [New York: Basic Books, 1992]).
19. On this, see the closing passages of my Introduction to Film as Philosophy (edited by R. Read and J. Goodenough [New York: Palgrave-Macmillan, 2005]). See also Cora Diamond's work on Wittgenstein (for example in her The Realistic Spirit [Boston: MIT Press, 1991]). And Chapter 3, above.
20. See e.g. Philosophical Investigations 119.
21. And what is climate denial based on, if not denial of limits, denial of death? A refusal to look in the face that we are at present, as a species, committing slow ecocide/suicide, in our 4x4s, in our industrial-growth capitalism, and so forth. See my "Growth and death," *One World Column*, 17 Jul. 2010, http://oneworldcolumn.blogspot.com/2010/07/growth-and-death.html. *Melancholia* might be profitably read alongside a film which also appeared in 2011 and with which it has sometimes been compared: *Take Shelter*. The latter, like *Melancholia*, might appear superficially to support climate denialism; read more deeply, the opposite is true. These films in fact accustom us painfully to come to terms with the unbelievable truth, which is as hard to listen to as the truth that we as individuals are going to die: the truth of the very real possibility of climate apocalypse destroying our collective existence. The protagonist of *Take Shelter* is made out by those around him to be a weirdo—for thinking that there is climate chaos coming. He is the one who cannot face reality, those around him insist: but in fact it is they/us who cannot or will not face reality. We have to de-retreat: we have to give up the impulse to take shelter in denial. Yet this is very hard to do, because the implications of facing up to the threat facing us all are so vast that we would rather find an excuse, grant ourselves a "reprieve."
22. Although I would suggest that most scientists in the world today are tacitly encouraging us not to worry too much about what we are doing to our world. Very few are really sounding the alarm.
23. Cf. www.greenhousethinktank.org/facing-up-to-climate-reality.html, for explication of what facing up to climate reality means.
24. I attempt to rise to this challenge in my paper, jointly written with Phil Hutchinson, "Practising Pragmatist-Wittgensteinianism," in Alan Malachowski (ed.), Cambridge Companion to Pragmatism (Cambridge: Cambridge University Press, 2013). Hutchinson and I argue in this piece that we humans should feel shame at the future we are currently co-creating, and even perhaps experience pre-emptively a hypothetical "grief" for the children and future people we are threatening—for these might well be just the emotions needed to prompt us collectively to stop the extinction of civilization. Cf. my discussion on the same theme in the Conclusion to the present work.
25. See my article on this, "Emergency Talk," *The Guardian*, Tuesday, 13 Nov. 2007, www.guardian.co.uk/commentisfree/2007/nov/13/emergencytalk. (Compare the

complete failure to develop a discourse adequately addressing the appalling situation within *NLMG*, as discussed in the previous chapter. Cf. also my discussion of "Hunters in the snow", later in this chapter.)

26. Note the way in which, in the film, the "fly-by" makes the atmosphere less life-giving (less breathable). The same move, notably, is made in *Avatar*, where humans are unable to breathe the CO_2-heavy atmosphere of the extra planet in that film, Pandora. The same occurs in *Gravity*, at the pivotal moments in the film (see Chapter 5), and in *Take Shelter*, with the gas masks that take centre-stage in one of the most important and visually stunning scenes in that film. *Melancholia* is every bit as helpfully obsessed as those three films are with breathing, only a little more subtly than they are; yet often in *Melancholia* the camera dwells on someone breathing, or struggling to breathe.
27. It may seem that the extreme deadening characteristic of much severe depression is a counter-example to this (as for instance in the partial-derealization Justine suffers on eating the meatloaf that "tastes like ashes"). I believe, on the contrary, that that is a defence against it: one avoids thinking what the worst thing that can happen is only by deadening oneself to it and to everything else; by deadening, flattening out life to nothing. To ashes. This is the ultimate attraction of "derealization" (and similarly of "depersonalization").
28. I explore their attractions in detail in Chapter 6, below, and also in section 2.3 of my Wittgenstein Among the Sciences, edited by Simon Summers (Farnham: Ashgate, 2012).
29. I argue this case in detail in my discussion of *Lord of the Rings*. See also my "Extreme aversive emotions: A Wittgensteinian approach to dread," in Ylva Gustafsson, Camilla Kronqvist and Michael McEachrane (eds.), Emotions and Understanding: Wittgensteinian Perspectives (Palgrave-Macmillan, 2009).
30. For further explication, see the *Lord of the Rings*, Chapter 6.
31. Thanks to Catherine Rowett for this point.
32. Cf. "How to destroy the Earth," *Geocide*, http://qntm.org/destroy! More seriously, think of the potentially tempting, if unappetizing, challenges posed by Benatar's Better Never to Have Been (Oxford: Oxford University Press, 2006), and the Voluntary Human Extinction Movement at http://vhemt.org/. Thanks to Kristof Bodnar for these references.
33. And of most horror movies, and of the new unpleasant extreme-crime genre, post-*The Girl with the Dragon Tattoo*; and more besides. Also, of eschatology: the not un-widespread emotional attachment to apocalypse narratives can of course be traced back to the mediaeval world and further.
34. In Sartre's sense; see his eponymous 1945 book, the second in the Roads to Freedom trilogy. (Originally published as "Le sursis", Paris: Gallimard, 1945).
35. The movement of thought and feeling here mirrors (and inverts) closely that depicted in Kazantzakis's/Scorsese's *The Last Temptation of Christ* (1988), as Christ overcomes the temptation to want his life to go on, to renounce the cross.
36. See T. Roszak, Mary E. Gomes and Allen D. Kanner (eds.), Ecopsychology: Restoring the Earth, Healing the Mind (San Francisco: Sierra Club Books, 1995).
37. Whereas, one isn't trapped somewhere if one is reconciled with it. Justine isn't trapped anywhere, at the end of the film.
38. As in the notorious scene of Justine naked, referred to in n.14, above.
39. Such moments of starting to come to life again are sometimes the most dangerous for a seriously depressed person. That is the point at which some depressed people kill themselves, because they cannot bear it when they find themselves starting to hope again (because then they risk being disappointed; they are no longer "in control"). On the possible reading of the film in which Part 2 is Justine's (or Claire's) fantasy of world catastrophe, of one's/the world ending, then, it would not be surprising that her (and everyone's) death comes soon after such

a moment. Our sharing of Justine's will-to-the-death-of-all, our wanting Melancholia to hit, is our own yielding to the attractions of the depressive vision. If we end the struggle and give up seeking to save our planet's threatened ecology, then we do not hurt so much.

40. Claire's words, of the plan for them to die while drinking wine on the terrace together, and of her hope for Justine to go along with that plan, are "That would make me happy." But, nothing *makes* one happy.
41. She "models," at last, the task of the viewer.
42. See Wittgenstein's Philosophical Occasions (Indianapolis: Hackett, 1993), p. 42.
43. Heideggerian being-towards-death authentically is sometimes referred to by Heidegger as "resoluteness" or "resolution." The reader may notice an interesting connection, perhaps more than merely verbal, between this and the "resolute" reading of Wittgenstein (aka the "therapeutic" reading). This connection might be explored for instance with regard to the broadly Heideggerian kind of reading of Wittgenstein's Tractatus that has been essayed by Eli Friedlander, Signs of Sense: Reading Wittgenstein's Tractatus (Cambridge, MA: Harvard University Press, 2001).
44. Thus in the end the problem deliberately offered one by this film, like so much in philosophy (so much in this book), is, just as Wittgenstein held, a problem of the will much more than of the intellect. The problem with *Melancholia*'s critics is that they do not want to understand what the film is calling us to, because to hear that call would open them to the need to act, and to the risk of disappointment. In other words, and in a savage but unavoidable irony: the film's critics are willing themselves to remain in the depressive position, deliberately rejecting the transformation that the film offers one in its final 30 minutes or so.
45. The films highlights the prevalent excessive calmness in the face of impending catastrophe, our ability to somehow be chilled as we gradually voluntarily commit the planet to very dangerous climate change, and perhaps to the Venusian scenario which James Hansen now fears (see e.g. James Hansen, "Making things clearer: Exaggeration, jumping the gun and the Venus syndrome," 15 Apr. 2013, www.columbia.edu/~jeh1/mailings/2013/20130415_Exaggerations.pdf). This ought to be distinguished from the needfulness of calm in the face of mortality and morbidity, as discussed helpfully in Simon Critchley's essay on *The Thin Red Line*, in my 2005 collection.
46. In this way, it bears a close family resemblance to *Dancer in the Dark*, a masterful anti-musical, a musical offering a very critical reading of the genre of musicals. *Melancholia* is a disaster film that critically reads the desires of the viewer of a disaster film. The resemblance between the two films is close, because *Dancer in the Dark* too played directly with the question of just how bad things could get. Each song resulted in things reaching a new low.
47. Cf. on this the logic of my www.thelondoneconomic.com/opinion/climate-change-once-we-no-longer-deny-it-then-we-just-might-have-the-will-to-try-drastically-to-change-course/14/03/.
48. Any reader in doubt as to von Trier's own strong interest in film as a therapeutic experience should watch or re-watch his *The Five Obstructions*.
49. The resurrection, as one might put it: look again at the image, reminiscent of Christ on the cross, of Justine in the film's prelude. In this vein, Justine appears as an inheritor of Bess, von Trier's Christ-crossed-with-Mary-Magdalen figure from *Breaking the Waves*. There we see Bess pushing her bike up the hill like a cross, having been forsaken by her mother, making the phrase "Christ on a bike" take on for the first time ever a serious meaning.
50. This point is discussed at length in Louis Sass's Madness and Modernism.
51. See https://onabutterflywingblog.wordpress.com/2016/11/03/the-hunters-in-the-snow-in-solaris/.

52. Compare p. 75 of Stanislaw Lem's book, the original version of Solaris (London: Bloomsbury, 1970 [2003]), for a nice example (one of many) of the story's engagement with and interest in denial. (Compare also the intelligently depicted moment in *Melancholia* of Claire's pausing and drawing breath before measuring with the tool Leo had created the progress of Melancholia towards [rather than, as should have been occurring, away from] Earth. She waits because she knows the evidence of her eyes—that Melancholia is approaching again, that the Dance of Death is real, that the end is nigh—but doesn't want to believe it. She allows herself a last moment of hope and denial, a pause to gather herself.)
53. The Great Derangement (London: Penguin, 2016), p. 42.
54. www.youtube.com/watch?v=FcglyhUre4w&feature=player_embedded.
55. See the detail in the image, here: https://52moviesinacollegetown.wordpress.com/2014/06/30/hunters-in-the-snow-2/.
56. See https://vimeo.com/38654145.
57. *Tristan and Isolde* is of course about two lovers (like the Earth and the Melancholia, locked—fatally attracted—in a gravitational dance of death), and their love affair leads to a catastrophe. (Tarkovsky's *Solaris* too is of course centred around a tragic love-story, that of Hari and Kris (Kelvin).)
58. Though there are "cracks" in this fantasy of his home house: the running water from the ceiling (water, because perhaps that is as it were the raw material with which Solaris works). One cannot live in that house. This is the ultimate tragedy of the film: Even a fantasy, an illusion that is produced by a higher being (by Solaris)—and that Kelvin perhaps even half-consciously accepts as such?—is flawed. There is an illusion within any illusion, so to speak. Or: no illusion is perfect. This is the actual lived experience of virtually all psychosis, for worse or better. (Cf. my discussion of *LYiM* in Chapter 2.)
59. See Chapter 5. However, Snaut's crucial remark at 1:53.58, "We have no interest in conquering any cosmos. We want to extend the Earth to the borders of the cosmos." dovetails well with the Arendtianism of *Gravity*. Snaut nails the tendency of humanity to practice a kind of collective solipsism, in the name of 'progress'.
60. As opposed to an acceptance of facticity, and then perhaps a determination from now on to change things.
61. "*Solaris*: Or, Do we really want to make contact?", http://thinkingfilmcollective.blogspot.co.uk/2014/07/solaris-or-do-we-really-want-to-make.html. Johnston captures admirably the way in which the film evokes a profound otherness in Solaris; and, as I read it, the way in which, from the very start, on a second or later viewing, *Solaris* offers us a sense of the profound otherness of (water and of) plant life, on our own planet. (Such otherness might of course itself be compared to the gulf between ordinary experience and psychotic experience; a gulf which, especially in Chapter 6, I am trying to diminish a little.)
62. Many thanks to Jakub Macha for discussion that has influenced my understanding of Tarkovsky's use of the Bruegel painting, and more generally has I think inflected my take on *Solaris* much for the better.

5 Gravity's Arc

Or *Gravity: A Space Odyssey*

> The true miracle is not walking on water or walking on air, but simply walking on this earth.
>
> —Thich Nhat Hanh

In the opening passage of *The Human Condition*, Hannah Arendt comments on the significance of the launch of the *Sputnik*, the first artificial satellite, into space. Her words are worth quoting at length:

> In 1957, an *earth-born* object made by man was launched into the universe, where for some weeks it circled the earth according to the same laws of *gravitation* that swing and keep in motion the celestial bodies—the sun, the moon, and the stars. . . .
>
> This event, second in importance to no other, not even to the splitting of the atom, would have been greeted with unmitigated joy if it had not been for the uncomfortable military and political circumstances attending it. But, curiously enough, this joy was not triumphal; it was not pride or awe at the tremendousness of human power and mastery which filled the hearts of men, who now, when they looked up from the earth towards the skies, could behold there a thing of their own making. The immediate reaction, expressed on the spur of the moment, was *relief about the first "step toward escape from men's imprisonment to the earth."* And this strange statement, far from being the accidental slip of some American reporter, unwittingly echoed the extraordinary line which, more than twenty years ago, had been carved on the funeral obelisk for one of Russia's great scientists: "Mankind will not remain bound to the earth forever." . . .
>
> The banality of the statement should not make us overlook how extraordinary in fact it was; for although Christians have spoken of the earth as a vale of tears and philosophers have looked upon their body as a prison of mind or soul, nobody in the history of mankind has ever conceived of the earth as a prison for men's bodies or shown such eagerness to go literally from here to the moon. Should the emancipation and

> secularization of the modern age, which began with a turning-away, not necessarily from God, but from a god who was the Father of men in heaven, end with an even more fateful *repudiation of an Earth who was the Mother of all living creatures* under the sky?[1]

These words anticipate a tendency that has grown stronger since they were written, and that Cuarón's *Gravity* starkly *brings home* to us: the deeply strange yet seemingly widely compelling desire to "escape" from the Earth. Arendt begins by taking an explicit interest in the pseudo-escape from Earth, or at least from full enfoldment in its gravitational field, that, for the first time in human history, became possible, at the dawn of the "space age." The power we now have to fire objects outside of the Earth's effective gravitational pull, I am going to argue, has fooled us into thinking that we have the power to truly move *subjects* out of that pull.

Arendt focuses on the profound danger—of giving up on the Earth—implicit in the entering of humanity into space. She dissects without mercy our hubristic, supra-Promethean, Icarus-like desire to fly away, to leave our Mother behind. She forces us to think what this desire means, and what, by way of an Earth-oriented "spiritual" perspective, we might have to recapture in order to overcome it.

All these elements are present in the alluring and frightening experience that *Gravity* offers to the viewer, an exploration of what it would be like actually to go through the possibility that Arendt warns us of.

Gravity is a very unusual Hollywood blockbuster. It starts not with some sci-fi premise, but with a fairly matter-of-fact explanation being given us on the screen of why "life in space is impossible." It then unfolds what is basically a structurally-simple story about a single character, cut off from the rest of humanity for most of the duration of the film by virtue of being not Earth-bound but space-bound. This character does not have the "gender-neutral" maleness implicit in some of Arendt's words above and challenged implicitly by Arendt's invocation of Mother Earth: this character is a woman,[2] albeit a woman in "a man's world," or rather, in a man's space. The film's script itself seems to implicitly acknowledge that its focus on a female character is unusual, that astronauts are implicitly "coded" male. The character is called Ryan Stone, because, she explains to experienced astronaut Matthew Kowalski,[3] her parents wanted a boy. In other words: the woman at the centre of this movie is taking up a place usually reserved for men. She may have been unwanted—but (t)here she is.

The fact that Ryan Stone is female is crucial for the story because it makes it possible for her once to have given birth to a child. She is (or rather, was) a mother. This allows the film to focus on the most primal bond between two human beings—that between mother and child—and on the sense of loss that comes with the severance of that bond. At the same time, *Gravity*'s dialogue, like Arendt's musing, refers to our planet as "Mother Earth," so that Stone, cut off from other people, appears in turn as that mother's daughter who is about to be lost. We can go even further: Earth is a giant rock in space, and the woman at the centre of this story is a "stone" circling around it. (If she were to die up there, she would after a while be as inert and cold as stone.) The intimate character study and the spectacular space adventure (the micro- and mesocosms, as we might put it) are thus presented in close parallel with each other.

Let's take a look at the character study first. Ryan Stone's daughter Sarah died in an accident when she was four years old, and Stone has never been able to process that loss. In some ways her life has been suspended ever since. (Could we say that she has almost turned to stone, emotionally speaking?) She says that since Sarah's death her life has consisted of nothing but work (as a doctor in a hospital) and driving from and to work (while listening to music—never talk—which fills the void surrounding her). On two occasions during the film (in conversation with Kowalski at the beginning and in a monologue towards the end), Stone states explicitly that she does not have any intimate bonds with anyone.[4] There appears to be no boyfriend, nor does she seem to be close to the father of her child. She barely mentions her parents, and when she does so her mentions of them are negative; nor does she even appear to have any close friends. Perhaps she intentionally keeps her distance from people because she does not want to experience another devastating grief. Here then we have someone with no felt bonds to other persons—almost an icon of our absurdly "individualistic" pseudo-society[5]—who ends up risking losing the bond with the Earth as our dwelling place. And after all: What better way could there be to keep one's distance from other people than to go into space? Stone hints at this motivation when she responds to Kowalski's question about what she likes most about space with "Silence," that is, one presumes, the absence of the noises made by human beings (rather than the absence of the sounds of the natural world, although, as we will see, on some level she might long for the absolute silence of death). Of course, at this point, there is no silence, because she is talking to Kowalski, and even when he is silent, the somewhat-annoying tinny music he listens to can be heard. There is a tension, then, between Stone's desire for silence (early in the film, we see she is not keen on Kowalski's verbal burbling) and her need nevertheless for verbal communication (and, sometimes, music). That need—for the connection with others that verbal interaction represents and creates—is something that becomes clearer as the action of the film proceeds. (It may be relevant here to think of language as a bond with the other, perhaps the last bond she has.)

There might be an intriguing parallel to all this in Kowalski's story: he is a raconteur in space, relaying tales about life on Earth, which revolve around failed human connections (an ex-wife who cheated on him, a Mardi Gras date that is over before it even begins). His ambition in life is to go on the longest space walk in history, floating around the Earth all on his own. And he gets to realize this ambition. The circumstances are tragic, but also very slightly ambiguous: he has saved Stone after a terrible accident in space, and she ends up holding on to a tether that prevents him from spinning off into space and death. He argues that she won't be able to pull him in because her own ties to the space shuttle are too tenuous; instead he would pull her with him into space . . . unless he severs their bond, which he does.

It is a crucial moment. Ryan Stone may have gone to space to keep her distance from people and to find silence; if that is the case, she gets more than she bargained for. The accident in space cuts off all communication with Earth and kills all crew members of the space shuttle except for her and Kowalski—who now leaves her behind (although he will be able to speak with her for a little while longer). At the same time, Kowalski's noble sacrifice points to a certain willingness to cut his links with humanity for good—and to die all alone. Importantly, Stone refuses for a while (in fact, for longer even than she realizes, it will turn out) to accept his apparently inevitable loss.

The film does not fill in all the psychological details, leaving you to do so. But it does suggest that space—and eventually death—is a void that some people, especially those who have lost loved ones or who lack human connectivity, might fantasize escape into so as to prevent further suffering arising from their bonds to others. Stone herself suggests this when she later imagines Kowalski's magical return which, in a pivotal, powerfully filmed scene that one experiences largely from Stone's point of view (and which makes a second watching of the film totally different from a first watching), is not initially signalled as her fantasy, but is eventually revealed to be just that. In this fantasy, "Kowalski" gently accuses her of wanting an easy way out of life's struggles by giving up the fight to survive, instead peacefully going to sleep until she is poisoned by carbon monoxide. This is indeed what Stone is trying to do—it is also, someone might venture, what Kowalski has perhaps already done.[6]

Stone's will to live is revived by her fantasy of Kowalski's return. On some level, perhaps, this fantasy establishes the kind of link to another person which, she says, she no longer has on Earth. She feels connected to Kowalski who (in her fantasy) knows her well enough to identify her wish to die and who cares about her enough to confront her about it so as to change her mind. At the same time, of course, this very fantasy ensures that, at least in her mind, in her soul, *Kowalski is still alive*; death is not the end.[7] (I will return to this point.)

The scene culminates in the following "exchange" between Ryan Stone and the "Kowalski" who, it turns out immediately at the end of this "conversation," is actually in her mind:

"KOWALSKI": Listen, do you wanna go back, or do you wanna stay here? *[Ryan says nothing]* I get it, it's nice up here. You can just . . . *[turns off the lights]* shut down all the systems . . . Turn out all the lights. . . . Just close your eyes and tune out everybody. There's nobody up here that can hurt you. It's safe.[8] I mean what's the point of going on, what's the point of living? Your *kid* died. Doesn't get any rougher than that. But still, it's a matter of what you do now. . . .

You gotta plant both your feet on the ground and start living life.

RYAN: How did you get here?
"KOWALSKI": I'm telling you, it's a hell of a story. Hey, Ryan?
RYAN: What?
"KOWALSKI": It's time to go home.

It's a hell of a story—because actually it is *Ryan's* story, the story we have been watching. And now, some part of her deep inside is telling her that the bond with Earth isn't over yet. That it's time to become grounded again, in the best sense of that word.

Just *prior* to that, Stone had been reminded of such bonds when she established radio contact with a man on Earth—not someone from the space centre in Houston, as she had hoped, but a radio amateur who speaks in a language unknown to her, but manages to communicate something important anyway by bringing a dog's voice to the microphone and then (closer still) a baby. Stone gets moved when she hears him singing to the baby, because it reminds her of her singing to Sarah. (She also seems to be reminded in the scene of something important and connective that she feels her parents omitted to teach her.)

Here is the key part of her monologue that closes *that* scene:

> "I'm gonna die, Aningaaq.
> I know, we're all gonna die.
> Everybody knows that.
> But I'm gonna die *today*.[9]
> Funny, that.
> You know, to *know* . . .
> . . .
> Nobody will mourn for me,
> no one will pray for my soul.
> Will you mourn for me?
> Will you say a prayer for me?
> Or is it too late?
> I mean, I'd say one for myself,
> but I've never prayed in my life, so . . .
> Nobody ever taught me how.
> Nobody ever taught me how.
> [MAN SPEAKING]
> [BABY CRYING OVER RADIO]
> A baby.
> [MAN CONTINUES SPEAKING]
> [MAN SINGING IN FOREIGN LANGUAGE]
> There's a baby with you, huh?
> Is that a lullaby you're singing?
> [BABY COOING]
> That's so sweet.

I used to sing to my baby.
I hope I see her soon.
[MAN CONTINUES SINGING]
[ALARM WAILING]
That's nice, Aningaaq.
Keep singing, just like that.
Sing me to sleep, and I'll sleep.
Keep singing. And sing and sing."

She'll be sung to, as she falls "asleep." As she speaks this somewhat self-pitying "final" speech, a tear floats towards us: in 3-D, it is more marvellous than most of the (amazing) special effects in the film; it is a tiny globe, almost a whole world . . . of sorrow.

But just possibly, too, it is that connection with a dog and then a baby that means that, as she drifts off for the last time, something in her, as it turns out, will *pull her back*. Something deep will give her one more shot at living. In a way that is not stuck in the past, or isolated, but grounded.

Gravity, then, *deals with* grief. A link embedded in the language, where the Latin root of our word "grief" is the same as that for our word "gravity." *Gravis* is the common root of "gravity" (also in the sense of "heaviness"; and, strikingly, of pregnancy),[10] and "grief." Grief and gravity, in our historical subconscious, are the same thing: the grave, the heavy, that pulls us down and grounds us—either in the sense of giving us a ground (as in "I feel really grounded, it's good"), or in the (opposite) sense of preventing us from moving on (as in "The vehicle is grounded! I can't move!"). Grief, I would argue, centrally concerns a refusal to allow that the world no longer includes the dead person.[11] Both phenomenologically (i.e. in terms of our lived experience) and logically (i.e. conceptually),[12] grief is the pain of a ruptured life-world. Grief is the lived refusal to too-quickly accept that someone important has been taken from us. As that person was a constitutive element of our world, an over-hasty acceptance of their exit would mean that we were not really denizens of that world, but merely observers of it, at best merely passing through rather than inhabiting it.

Grief is ungainsayably cognitive. For obviously it makes sense, it is *rational*, to have a world; which means: to care about those in it, those co-constituting it. Indeed, I would suggest that grief is essential to our humanity. One would have to be some kind of inhuman monster, and/or disabled in a profound way, not to feel grief under appropriate circumstances. However, like everything rational, grief can be pathological if it becomes overbearing or permanent, turning into depression. Stone is letting go of that depression, at last, when she overcomes her desire for death and realizes that, due to their shared experiences, their influence, their values, her daughter (and also Kowalski) lives on in her. Thus, grief—and *Gravity*—is a forceful reminder of the "fact" (that is deeper than any mere fact)[13] that we are not separate from one another, but always connected, even beyond death. (In this sense, to vary

William Faulkner: The dead aren't dead. They're not even past. They're present; when we are truly alive, when we dare to live.) The film is thus about grief, and its overcoming, as the acceptance of (inter)dependency, rather than striving for independence (a striving which is so closely associated with American history and culture). Interdependence—and none more so than the relationship between mother and child—makes us vulnerable but it also ensures that we live on in each other. Grief concerns *how we are not separate from one another*. Kowalski is still there, part of her, after he has gone. We are part of one continent. Stone is not alone, even in space. In space, even if no one can hear you wail, you are still not alone. Stone is connected to the man she speaks to on the radio on Earth, to his dog (we are animals, and close to animals), to his baby. She is not severed even from Kowalski, nor from her daughter (even though they are dead—and, as one finds in the spin-off film, the dog that she has just heard barking too is about to die).[14] She is connected, we might venture, to you and I; and we to each other.

Thus, crucially, we can say that *Gravity* shows how grief need not be felt as a regrettable weight upon one. Milan Kundera famously asks in *The Unbearable Lightness of Being*, "But is heaviness truly deplorable and lightness splendid?"[15] The grave can be a route back to life, life as part of the continent, part of the planet, for one prepared to fully face it, embrace it, and carry on. Carry on as and with others, including others who are dead (and, I would add, others who are yet even to be conceived: the unborn future generations who depend utterly upon us to steward the Earth). In zero gravity, in space, one might naturally hope to escape (such) weight. But there is no escape: and this is a good thing. The way through grief is *through* it, and through the wonderful acceptance, Nietzsche-style/Zarathustra-style, that one would not for all the world have had the world lack the people who died, even if the cost was that one had to grieve for them.

We can take this connection one stage further still. Nietzsche's Zarathustra famously opposes "the spirit of gravity." Stone in the film is both constantly and occasionally consumed by grief—constantly over her lost child, and on occasion over Kowalski's loss and over her own surely soon-to-be-lost life. Grief moves naturally towards its (ever-absent) object. Stone is, for most of the film, *grave*, and, when she decides briefly to embrace and hurry up her death, she seeks to meet the object (the *subject*, Sarah) of her grief (in Heaven); *but*, by then working through her grief in a context in which she is confronted with her ownmost death,[16] she escapes the grave that seemingly awaited her in space. And she even manages to enjoy the process, to enjoy the moment come what may: thus her thrill at the bumpy ride home. Through the grace of luck and of Kowalski (and her memory of him) and of Earth's beauty (that he turned her and our attention to) *pulling* her home: *through* gravity, through allowing and accepting the physical force, weight.[17] Earth is our home: it can even mother us as we integrate our pain at failing to be adequate parents/stewards and adopt a healthier relationship to our past and future. This film offers a subtler way of achieving such health than Nietzsche's Zarathustra does, with his sometimes heavy-handed insistence on joy and his failure to adequately relate us to one another and to our planetary home.

And it offers in the process then precisely a Kunderian reworking[18] of the traditional polarity of gravity and grace. Vast cultural and historical forces and hegemonies assume that to be light, to be spirit, to be "graceful," is preferable to being heavy, earth-bound, subject to gravity. Christianity is suffused with and has buttressed this hegemony. *Gravity* by contrast joins Philip Pullman (in *His Dark Materials*) and a small band of others who find our fleshly, earthly existence a truly splendid thing, the true home of spirit. We need to learn to "plant both [y]our feet on the ground," as at one pivotal point (as we saw above) the spirit of Kowalski tells Stone.

We are told again and again by our culture and by most science fiction that to boldly go where no man has gone before is noble, whereas to be at / go back home is banal. But the elixir that (Kowalski and) Stone bring back to us from space is that this is all wrong. The true hero's journey now is to return home: to really feel this as our home, and to live well *here*. (In this connection, it is worth being encouraged by the thought that the greatest real-life space adventure story that we know is the incredible story of Apollo 13—a story precisely of managing to come home from the utter inhospitability of space and despite the tenuousness of human technology.) That is what we would do, if we loved life. Traditional stories praising angels who float lighter than light in the heavens are perhaps subtly a call for us to feel inadequate, hate life, and want to quit the Earth.[19]

Gravity adds a dimension to its renunciation of depression and its plea for life by making palpable the sheer excitement and wonder that life can generate. Right from the beginning of the film, we find ourselves moving around in space high above the Earth, enjoying breathtaking vistas but also soon experiencing extreme danger and utterly disorienting movement. Initially, the film's largely computer-generated imagery creates the illusion of a camera's continuous movement around spacecraft and bodies, and also into the very positions from which characters view the world around them. The deployment of director Alfonso Cuarón's trademark ultra-long tracking and panning shots at the start of *Gravity* is a technical tour de force, which may draw attention to these shots own virtuosity, but also adds to the film's thematic concern with the connectedness of inside and outside, character study and space adventure (and ecological parable), film and awakening (including from film).

Most of the film, especially in its early stages, can in fact be regarded as composed of tracking shots, "long takes." Let us explore the second extraordinary tracking shot in some detail, since it can function as a key to the film. The second long tracking shot in *Gravity* becomes genuinely extraordinary and meaningful as, in the course of the disaster (of the radiating space-pollution) unfolding, it begins to become a point-of-view shot. That is: we very gradually, in a drawn-out process, enter into Stone's point of view as she experiences the disaster of the impact of the space debris. This process begins in earnest at 12.45. At 13.48, we enter inside her helmet. At 14.00 we start to look out from her helmet, and then from her eyes—we are only now literally occupying her point of view, as she spins helplessly and gazes out, unagential and petrified. Interestingly, at this moment, right amidst the unfolding disaster, we first start to notice "the Earth is on fire," "alight" with man-made illumination. By 14.50 we've moved back out of her helmet again. At 15.30, we resume being independent of her, seeing her spin off into deeper space. All of this, in the felt-reality provided especially by 3-D, is a powerful experience. We are immersed in what it actually is to be losing one's connection with Earth and other human beings (and by extension, what it is to retreat from connection with others voluntarily, by contrast and comparison with involuntarily being cut off from them); what it actually is to be living somewhere, temporarily, where life is impossible; and what it is to be able to see what is happening to us-and-the-planet.

I have only dwelt in the above paragraph on the visuals; the intelligent way in which this long scene is filmed in terms of its audio enhances the effect yet further, especially by means of sound and the absence of sound and the fading out and in of sound. This aural effect is one of being cut off from others; what is so clever about its execution is that we tend not to even notice that it is happening.[20]

At 18.50, we enter into Stone's point of view again—into her helmet. The film is cementing the way in which, as I opened by saying, it is for much of its duration a film with one live character only. One life, like our very own.

Spectacular views of Earth and space, and marvellously slow and dextrously rapid camera movement, provide us viewers with a visceral experience. As first Kowalski, and then much later Stone, says: "It's a hell of a ride!"[21] "Ride" here initially refers to space travel (and by extension, back to driving, which, as you'll recall, Stone does pointlessly on Earth, immersed as she is in her depressive grief), but, more generally, it takes on the meaning of human life (and also the film we are watching). The film takes us on a ride which is meant to remind us of the thrill of being alive; something only possible on or very near Earth. This continues for most of the story, which moves from exterior space to the interiors of various spacecraft until, finally, Stone plunges back to Earth in a small capsule.

Before we get to this point, the film examines the ambiguities of space exploration. Stone is in space because a device she developed for use in hospitals can also be used in the Hubble space telescope which, we are told, is designed to reach out to, and gather information from, "the edge of the universe": exploring and healing the human body is connected to the exploration of the whole universe; looking inward and looking outward are two sides of the same coin.

The film never mentions the physical exploration of outer space, with manned and unmanned spacecraft escaping Earth's gravity altogether so as to go to the Moon and beyond. This is part of its realism, much greater than most of its predecessors as to the nature of life in space, which is likely to be virtually impossible for healthy human beings for periods longer than a few months, or at most years (we shall return to this point; if you see what I mean). Instead, in this film, people and their craft remain in Earth's orbit, which provides them with spectacular views of the planet's surface. Indeed, Kowalski's last words—while drifting off to his death in space—concern the beauty of Earth and thus, it is implied, of life, and they are spoken precisely so as to give Stone a reason to go on. He speaks of the beauty of the sun shining on the Ganges in the hope perhaps that this great beauty, together with Earth's gravity, will pull Stone home.

However, as already mentioned above, the view from space has another dimension. Where there is night on Earth, the artificial light resulting from human habitation looks like a slow-burning fire destroying everything in its way (like lava flowing off a volcano). In a tradition going back to the first widely disseminated pictures of the Earth in space (notably the ones known as *Earthrise* and *Blue Marble* from the late 1960s),[22] seeing the globe reveals both its beauty and its vulnerability. The Earth is on fire: threatened, as a home for life, by climate-changing, temperature-raising carbon emissions, the pollution spinning off from our compulsive, destructive "growth."

At the same time, near-Earth space is shown to be a would-be new habitat for humans, who fill it up with various spacecraft. Two permanent space stations (an international one and a Chinese one) are pioneering outposts of humanity, with, possibly, significant waves of human migration to follow so that we might imagine that, like all the continents of Earth before, space as well may be colonized. Yet this possibility, and more generally the human use or "development" of space, is by no means unproblematic. One specific reason why it is not unproblematic, explored explicitly in the film, is that with human habitation comes environmental destruction through new forms of pollution—even in space.[23] When a Russian rocket destroys one of the Russians' own satellites (a spy satellite with sensitive technology it would seem, which the Russians are apparently trying simply to retire), a chain reaction is triggered, whereby debris from the satellite slams into other spacecraft, creating more debris etc. This (a realistic potential scenario)[24] is the cause of the accident that kills all members of the space mission Stone belongs to—and also leads to the abandonment of the two space stations she flies to in search of an escape capsule. With accumulating space debris forever circling the Earth, humanity's colonization of near-Earth space has already begun to cancel itself out.

In this context, the film's title takes on a still-larger range of meanings. Most banally, one might say, the story concerns a serious, "grave" situation—Stone finding herself stranded in space as the lone survivor of an accident. The "gravity" of this situation is intensified precisely by the fact that any outside help would now have to overcome the pull of Earth's gravity so as to join her in orbit—and by the fact that some of the space debris is held in much the same orbit by Earth's gravity. Even if it was not extremely difficult in the first place to send a rocket to her rescue, such a rescue mission would be almost impossible due to the dangerous debris now circling the Earth.

The bond between Earth and this debris echoes perhaps the bond between Stone and her past via the "painful debris" of grief. The macrocosm and the microcosm, again. Macrocosmically, we are polluting Earth and even space. If we can come to terms with what we've done and with our past, maybe we can yet stop the future from hurtling out of control.

Complementing the pervasive imagery of tethers—tenuous yet vital links between people or between people and spacecraft—Stone's floating in space is the result precisely of being tethered to Earth by the planet's gravity. Rather than drifting off into outer space, she continues to be connected to Mother Earth by a kind of ethereal umbilical cord.

When Stone finally manages to find a spacecraft with which to return from her orbit to the planet's surface, gravity becomes a potentially deadly force. Gravity accelerates the plunging capsule so much that it almost burns in the atmosphere; and yet it is only the pull of gravity that can bring her home. Here we may be reminded of the trauma Stone has been trying to escape from: her daughter played at school and fell down, gravity, together with her own momentum, pulling her to the ground with such force that she broke her neck. At the end of the film, then, we are reminded of the deadliness of gravity and also, at the same time, of the fact that it is the basis of our lives. This reiterates, on another, global level, the central point I have made: the film's focus on grief serves to emphasize the fact that humans are dependent on each other, which makes them (us) both profoundly vulnerable and indestructible (because we outlive our own deaths, through others, through life going on). Similarly, the film's focus on gravity expresses our dependency on the Earth—it might seem to tie us down, but really it anchors us. It lifes us as well as giving us a kind of material afterlife, because eventually our bodies become earth.

Stone's return to Earth is presented in archetypal imagery. She confronts the four basic elements of old: the air of the atmosphere,[25] the fire that almost burns her capsule, the water of the sea into which the capsule falls, and the earth she crawls onto afterwards. There is also the vision of what appears to be virgin land, untouched by human habitation, a kind of paradise which Stone is allowed to (re-)enter—while the radio messages on the soundtrack have assured us that she is not in fact alone, that human company is on the way. *Gravity* thus depicts both the continuity of human connections and the promise of a new beginning—not just for Stone but also, just possibly, beginning with our watching a film like this, for humankind.[26] The film emphasizes the fact that she has to come very close to death before she can step on the Earth again; to be born again, first one has to die. As soon as she opens the capsule, it fills with water and sinks, and when she escapes from it, her space suit fills with water as well, dragging her down (she is indeed sinking like a stone). The technological devices that have protected her in space (capsule and suit) have to be abandoned for survival and a new beginning to become possible.

It is only after she has come very close to death one final time that Stone can finally make her way back to the surface and to land. In retrospect, the capsule filling with water and the sea appear both as deathtraps and as wombs from which she is born again. Her movement echoes the development of life on Earth—from water to land, and, on land, from crawling to walking. The film includes a reminder of this development by briefly focusing on a frog swimming upwards, like its amphibian ancestors that were the first to make the transition from water to land (and whose descendants are proving the most vulnerable of all to anthropogenic extinction). This frog in fact could be seen as *showing her the way* to life. Swiftly upon this, as she emerges, follows Stone's passionate embrace of mud, the mud that provided living space for the first creatures to emerge from the sea. She holds the mud, looks at it, laughs with happiness, relief, pleasure. While birds sing, she quietly intones "Thank you," looking down into the mud. This moment has been interpreted by some as an explicitly Christian gesture.[27] Possibly God(dess) is present in her mind in this scene, as possibly are the people who helped her get to this point (especially Kowalski, also the nameless radio amateur), who helped her to live. Better, one could think of her as thanking the very gravity that pulled her down, in all the senses of that word "gravity" that I have explored above. But best of all, most plausible, direct and parsimonious, is to take her here to be thanking the Earth itself, that was and is ever producing this gravity, and its fertile soil (earth) that is here manifested by the mud. After all, *that is what she is looking at* as she expresses her gratitude. Not, as one conventionally does when thanking God, at the heavens.[28]

Finally, there is Stone's struggle to get back on her feet (once again echoing or compressing millions of years of evolution). At the very end of the film, it takes every effort for her to stand up, finally towering almost majestically above the camera (which stays on the ground, looking up to her). It is hard to stand up and walk, as hard as it has been for Stone to overcome depression and return to life, return to the Earth. It is hard to accept and to cope having to come (and be) "down to Earth." And it is wonderful. She feels herself alive; we feel with her the sense of weight (just as before 3-D gave us the sense as if of weightlessness), the difficulty in standing up, *and* the wonder of being able to do so, and of walking on the Earth. Full of the wonder of it, striving to resist the pull of gravity to flatten her back down, she walks hesitantly forward. And the film ends. (And now, when one is ready, one can stand up, oneself.)

This final shot, importantly, contains a striking and lasting reminder of the presence of the camera, similar to the reflections and refractions of light on the camera's lens in numerous other shots, though even more "alienating" and drastic than them.[29] Here it is mud and water which have been splashed onto the lens by Stone's movements. As the camera is positioned on the ground, we can say that the dirt on the lens reminds us of its—and our—immersion in and reliance on mud, the same mud that Stone clawed into and cherished after having extracted herself from the water. The stained lens also reminds us, of course, of the very existence of the camera and the fact that we are watching a movie; in this, it is almost equivalent to the direct looks at the camera in the last frames of the action in both *2001: A Space Odyssey* and *Avatar*. As we shall see, both films revolve centrally, like *Gravity*, around the idea of rebirth (an astronaut being reborn as a "Star Child," a human being reborn as a Na'vi) and around the need, and the possibility, to gain a new perspective on the world we live in: the Star Child gazes at the Earth before it turns towards the camera, and Jake Sully, having woken up to the planet's beauty and vulnerability and having acknowledged adequately at last its inhabitants, abandons his human body so as to be able to live permanently in the (so hostile for humans) environment of Pandora. When the Star Child or Jake stare at the camera and, through it, at us, these films remind us that what is at stake in these stories is *our* perspective as well. Are we willing to see the world anew? And what might we be willing to do as a consequence of our new perspective? Might we, for instance, decide *not* to give up on the challenges we face today? I am talking now about us as individuals, us as part perhaps of a movement—and us as a species. *Gravity*'s ending addresses us in something of the same way, serving, like the ending of *2001* and of *Avatar*, as a call to action, by way of being first a call to contemplation.[30]

The "alienation effect" of the mud hitting the camera is, I would suggest, the film's final invitation to its viewers to heed its call, to think about what is offered in the experience of the film, to be *reminded*, in Wittgenstein's sense, of what one utterly knows but can be persuaded by ideology to forget: in this case, that life on Earth is worth saving, and that (for the foreseeable future) life for us is *only* possible on or near Earth. Thus the film seeks to transform us by returning us to life, to the awareness of the wonder of this life, and to the "fact" that being alive is a gift not to be discarded. *Gravity*'s space adventure ends with a renewed appreciation of many of the fundamentals of life on Earth—breathable air, fertile soil which is also the ground we can walk on, as well as great bodies of water that first nurtured life on this planet, and just as importantly, the human interconnectedness which sustains us. The space adventure in the film here stands in for the film itself, Stone's journey, her "odyssey," representing the actual or needful journey of every viewer: we let ourselves be taken into space by the film so as to return from this journey, just like her, with a renewed appreciation of our everyday surroundings, knowing them, and knowing our way about in them, perhaps, for the first time. This vision is offered us in the knowledge that without vision the people perish.

Thus the film can be helpfully summed up by saying that it, just like *The Lord of the Rings* (on which, see Chapter 6), is all about letting go and not letting go. And about letting oneself be attracted, moved etc.[31] Repeatedly, Stone has to let go; Kowalski tells her to, repeatedly. She has to let go of him

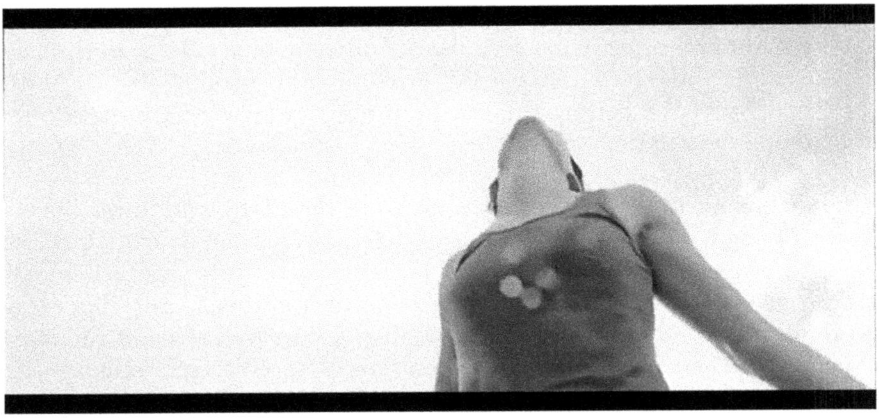

Figure 5.1 The finale of *Gravity*: Re-minding the audience.

once they are tethered so that he can re-orient them with their rocket; she has to let him go so that she can survive; and she has to let go of the idea of coming back for him, of saving him. But equally, she has to *not* let go: of Earth, of life, of the presence and reality of other living (and dead) beings, human and otherwise.[32] These two ideas are combined in one of Kowalski's final remarks to her: "Ryan, you're gonna have to learn to let go . . . I wanna hear you say you're gonna make it." She is causing herself anguish by the tightness with which she is clinging to her daughter's memory and to Kowalski's life; but she must cleave to life itself more tightly, if she is actually to survive. This is the path that Stone must tread—one of learning and enacting how and when to and not to "let go." This is another way of seeing what I discussed earlier in relation to grief and depression. As Kowalski's "ghost" says to her: "I geddit. There's nobody up here that can hurt you. It's safe." It feels safer to retreat than to take the risk of committing to life, to other beings, to hope. Even when the cost of such safety is death. The film offers us a path beyond such voluntary retreat in the face of our involuntary placement in appalling situations of isolation (and yet non-isolation) and of destruction.

Thus, the film engages in a profoundly ecological task with, hopefully, a call to action being able to be felt/heard, by viewers, as a result. Rather than leaving the Earth, we need to acknowledge our embeddedness in and indebtedness to it; and that requires protecting it as a space (a place) where life can flourish. To return to Arendt's words, which opened the chapter:

> The earth is the very quintessence of the human condition, and earthly nature, for all we know, may be unique in the universe in providing human beings with a habitat in which they can move and breathe without effort and without artifice. The human artifice of the world

> separates human existence from all mere animal environment, but life itself is outside this artificial world, and through life man remains related to all other living organisms. For some time now, a great many scientific endeavors have been directed towards making life also "artificial," towards cutting the last tie through which even man belongs among the children of nature.[33]

Gravity reconnects us—men and women—to the Earth, and to the life we share. Flying wonderfully in the face of the tradition of most sci-fi, it resists the pull to give up on the Earth, and in the most realistic spirit it reminds us of our profound—complete—connection to it.[34] It is a story that seeks to think philosophically and ecosophically, in part by means of the most remarkable virtual experiential inhabitation of specific extra-atmospheric points of view by the viewer.

Gravity offers viewers an opportunity to begin at last to question the hubris of technophilia in general and of space opera and of most sci-fi, and the disastrous alienation that Arendt means to interrogate. One could describe it as an Arendtian film, and more broadly perhaps as a Heideggerian or a Wittgensteinian one. But above all, like any major philosophical work, it is in the end doing something of its own. Something that, when it is done well, strikes us as if it is something that we always already knew.

Gravity's arc parallels precisely the arc imposed upon anybody (not quite any body, but: anybody) on Earth. To vary Newton: the film shows that and exactly (psycho-socially) how *who goes up must come down*. It escorts us through fantasies of escape back down to Earth. The phrase "down to Earth" has a wonderful double valence: it describes both the sad falling away from a fantasy (i.e. back down to Earth) and the happy state of being immune to such fantasy in the first place. The arc literally described by an object subject to gravity is described by Stone's post-Homeric voyage-and-return in the film: a psychical journey from the first sense of being "down to Earth" to the second. *Gravity*'s arc echoes precisely gravity itself, Stone's hero's journey and our own, away from Earth through fantasies of escape and then back to Earth, to realism and the acknowledgement of where we are, in a meaningful sense, "at home."

What/who goes up/away must come down/back IS the fundamental structure of narrative itself, of the hero's journey. *Gravity* is a marvellous whole of title, form, content, style: all these are the same, all one.

As the film explicitly has it: *launching is landing*. This is the lesson which Clooney's character's "ghost" teaches Bullock's character, and through which she saves herself and reunites with the real mother-ship, Earth. Leaving is arriving. Who goes up must come down. To (understand what it is to) love, you must understand what it is to lose; what (who!) is *missing* if someone doesn't return from a journey.[35] And when you do, but are no longer trapped in the past with/by such knowledge,[36] then you are ready to live again. Here, in the only place where life is possible.

***

Gravity strikingly replays many aspects of *2001: A Space Odyssey*. I have already mentioned some of the ways in which *Gravity*, the greatest of space odysseys in recent years, might be connected with or be seen to inherit important themes from its iconic predecessor, including, crucially, a closing "alienation effect" which I claim is also an *addressing* effect.

Consider some more:

- During the conferral of the two astronauts in *2001*, about their worry that the "infallible" (cf. unsinkable) A.I., HAL, is malfunctioning, Poole (who later dies) says, "I've got a bad feeling about it"; echoing him, in *Gravity*, George Clooney's character repeatedly says, "I've got a bad feeling about this mission";
- The dead astronaut Frank Poole's body drifts off into space, before being retrieved, movingly but pointlessly, like Sharif in *Gravity*;
- The tenacity with which the lone survivor of the Jupiter mission, David Bowman, clings to life and eventually returns home (after being reborn on his death bed);
- The way that the curve of the astronauts' helmets in *Gravity* echoes the curve of the Star Child's protective cocoon;
- Shots where Stone adopts a foetal position, and even slowly spins round like the foetal Star Child in *2001*;
- The emphasis on life and its vulnerability via the inhabitation of the inside of a space helmet in which breathing is heard particularly loudly;
- Point-of-view shots, widely and consequentially used in *Gravity*, and most famously, in *2001*, from HAL's perspective.

Above all perhaps (and I will return to this point), both films express an implicit critique of our un-Arendtian desire to escape our home and of our dangerous ever more complete dependence upon technology. *Gravity* and *2001* both apparently centre on a journey into space, but what really interests them is a *homecoming*. The "elixir" brought back by the hero(ine) is in the *return* of the voyager, symbolizing the possibility of awakening, of "rebirth."

So I'll dwell a little more on *2001* itself, *Gravity*'s great predecessor, in the light of my culminatory thoughts on *Gravity*. I want to focus on a plot mystery in *2001* which I think can be better understood once one has seen *Gravity* and thought adequately about our utter link to the Earth, literally the ground beneath our feet.

One of the key mysteries of *2001* concerns why HAL malfunctions. It seems to be a mystery to everyone. The astronauts (and not dissimilarly, viewers and critics) puzzle over why—*how*—HAL could (possibly) be malfunctioning, for there is an identical HAL on Earth, and *that* HAL is just fine. The mystery endures to the end of the film. This may, on the surface of it, seem to have little to do with what the story is essentially about. I think it has everything to do with it. For, of course, the two HALs are *not* identical. What the characters *in* the film almost wilfully fail to notice are two vital features of the situation, two contextual facts so obvious that one easily stares right through them. Curiously, these are facts that some philosophers also encourage us to (try to) ignore.

Philosopher Derek Parfit, in his famous "branch-line case" concerning personal identity, a little sci-fi story about one "you" on Earth and another "you" in space, concludes that we do not need to worry about one of these so long as there is another "you" elsewhere who is unharmed. "Psychological connectedness" (of someone with your past-self) is what matters, not your own continued existence.[37] *2001* asks us to consider two points which, if taken seriously, could contradict this view:

1. Interaction influences/constructs personal identity: on board the spacecraft HAL interacts conversationally with the astronauts, Bowman and Poole, whereas the "identical" HAL back at home does not. Moreover, think of the nature of the interactions: Bowman and Poole evidently do not (want to) think of HAL as any kind of "him," as *someone* who they have to *relate* to. They treat him merely as a servant. It is possible that HAL having to keep the secret that he is entrusted with, as he interacts with two beings who do not regard him as a "proper" being, becomes intolerable.[38] Either way the deep-space HAL is changed; he becomes who he is through conversational interactions.
2. The first point does not yet, however, provide a completely sufficient reason for HAL's breakdown/malfunction. The fact that HAL is treated as nothing but a machine combined with him having to withhold a major secret from the crew might amount to good reasons for some sense of alienation, or for some disturbance, but this does not fully explain the way in which HAL turns against his human colleagues so violently, and initiates a fight to the death. I believe that there is a further reason, unnoticed by critics,[39] for the grave difference that develops between the two HALs, which triggers the full-on breakdown of deep-space HAL. It is this simple fact, so obvious that one can overlook it: one HAL is on Earth, the other is in deep space. My hypothesis is that the HAL on board the spacecraft starts very slowly to go off the rails for the same reason that a human might: he is (virtually) alone, vulnerable, and (so) frightened.[40] HAL is further from Earth than any being has ever been (closer perhaps to the mysterious aliens underlying the story); and so, naturally, he is scared. The feeling of separation worsens because he has no support and no sense of himself as a being from among the humans he works for and from whom he has to keep a large secret. The human astronauts (brilliantly played by the two lead actors) come closer to being the ultimate professionals, seemingly unaffected by the extremity of their isolation. They control their emotions under

> the most extreme pressure.[41] Think of the long, silent scene where Bowman seeks, while wrestling with fury (and presumably also with fear?), to find a way of re-entering the spaceship, as HAL seeks to block him from doing so. The astronauts are, one might venture, closer to being robots emotionally (as we usually imagine robots) than HAL.[42] They all stand at the end of a desolately terrifying, long "supply line";[43] there is no resilience to their system. If one major thing goes wrong, they are done for. Yet at a conscious level, both astronauts are extremely, almost absurdly unemotional about their situation.[44] *They* seem almost *unconscious* of the grave danger they are in.

HAL is *responsible* for the mission, in all its tenuousness, which already makes him different from the "twin" HAL back on Earth.

My observations about the identity of space-HAL show how (like *Gravity*) *2001* could be seen as a reflection on our connection (or lack of it) with Planet/Mother Earth and the (anti-Arendtian) ease with which we ignore it/Her. This is represented in *2001* by the polarity between the (human) beings who should feel a connection most strongly, versus the computer which starts to feel what any sane human would feel in similar circumstances: unminding fear at his vulnerability. This fear is, on my reading, what triggers HAL's malfunctioning. What occurs is a vicious circle beginning with HAL's distant geographical placement, which sets him off on an entirely different psychical journey to the safe Earth-bound HAL. Fuelled perhaps by his only relationships (with the impassive crew members, whom he has to deceive) being inauthentic, deep-space HAL enters into conversations with the astronauts that lead to a descent into paranoia, both unjustified and justified, and finally to murder. We would not / cannot expect the Earth-bound HAL to continue the same as Space-bound HAL, for the latter inevitably takes part in a relationship with the astronauts that the former does not, in a context of extreme loneliness/isolation (and responsibility), accentuated only by their blankness. In this relationship/context, who HAL is develops and changes very consequentially. *Where* one is makes a difference to *who* one is. (In this context, Derek Parfit, in his famous "branch-line case" argument on personal identity, is rather like one of the humans in *2001*. One who does not understand why the two HALs are not simply interchangeable. Who does not understand, that is, how *where* one is can make a difference to *who* [or even *what*] one is.)

We can deepen our understanding of this non-interchangeability by taking even more seriously the "personal identity" of the far-apart HALs. For, as in most such sci-fi stories, there is a serious potential philosophical problem with the premise that we are supposed to grant, albeit not necessarily without hesitation: that HAL is a conscious, intelligent being. That it is possible to programme or somehow to "grow" a computer that really is worthy of being said to have "artificial intelligence." There are in fact a number of such problems, explored at length in Hubert Dreyfus's important

Heideggerian work, *What Computers Still Can't Do*.[45] These problems can be collected around the following, counter-Cartesian, thought: it is (to say the least) unclear that it can make sense for a *disembodied* "being" to have being, to be conscious, to be genuinely intelligent.[46]

The only possible defence I can see against this objection is to suggest that the beings in question are not disembodied. This defence obviously works in the case of the replicants in *Blade Runner*.[47] It is not clear, however, if it works in the case of HAL. But here is how it might work: if one considers the *spaceship* as HAL's *body*. After all, it is through the spaceship that HAL senses. There's plenty of evidence in the film of how HAL "inhabits" the whole ship, how it "is," in effect, him. (This is why, for the crucial conversation in which they discuss his future, Bowman and Poole have to lock themselves in a pod.) Some of the stranger angles from which events within the ship are filmed in fact make better sense when considered as point-of-view shots.

And now we can see how my argument is strengthened by the point concerning HAL's physical existence.[48] HAL's *location*, at the edge of knowledge, at the end of the longest possible supply line, makes a key difference, a difference that the protagonists in *2001*, in their human narrowness, fail to consider. But how much more of a difference it makes when one thinks of HAL not merely as a mechanical "spirit," but as embodied by this hulk of "living" metal that is vulnerably placed in deep space.

2001 is a magical work of art.[49] Like *Gravity*, it forces the viewer into a process of "therapeutic" questioning as to who we are, and what we can and can't do without, as (human) beings. And: what it is to be a *person*. Part of this process, I have suggested, involves coming to realize that who *HAL* is is affected by his history and his (extreme) situation in the same kind of way that humans, we utterly vulnerable creatures, are affected by such things.

2001 is an odyssey for our time—a hero's journey, *essentially* involving the return to where we began. In this way, it deeply mirrors *Gravity* (or rather: *Gravity* is legible as a kind of relatively "realistic" compressed re-writing of the same basic narrative).

2001 takes us on an odyssey through space, but also on an odyssey through our emotional and practical dependence upon technology. We start the film without technology. Then the first technology is envisaged as a weapon. So we start eating meat—eating other animals. That is, we become violent. Then we swiftly start killing each other. The film flips forward to roughly our current state, highlighting our virtually total dependence upon technology. This weakness "comes home to roost" in the microcosm of the film's plot, as our technology turns against us: and Bowman is forced to strip away his dependence upon HAL. That leads directly into the final phase of the film, wherein Bowman achieves a final awakening: one which seems to represent us *overcoming* our dependence on technology. The Star Child has no need for technology at all . . . just like our early predecessors

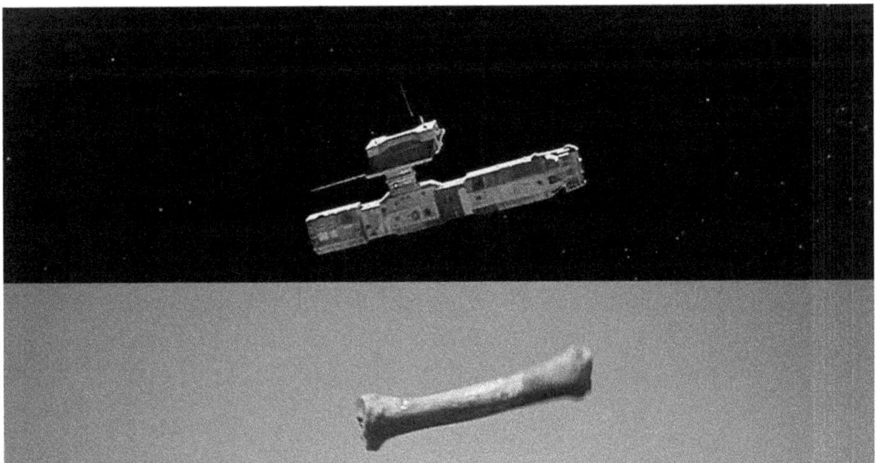

The 4,000,000-year cut

at the film's start. This rebirth, presaging Ryan Stone's less sensational but just as significant rebirth, is a return to where he started, but with much deeper understanding. As at the end of Eliot's *Four Quartets*: the end of all our exploring is to return to where—and in a certain sense to when—we began, and to know it for the first time, with a new sense of appreciation.[50] The hero/heroine is reborn, returning to share "the elixir of *life*" which, in Jungian terms, is the reward of individuation (which is not at all the same thing as individualism, quasi-solipsism), of becoming who one is.

HAL *is* our over-dependence on technology. As with the astronauts who he terminates, the flashing on-screen message "**Life functions critical**" just about sums up our collective state at this point in history.[51] *2001* and *Gravity* aim *inter alia* to wake us up (as Stone awoke from the permanent sleep she almost sent herself into, and from her permanent grief), so as to not send our living planet into a terminal permanent sleep.

Gravity is a space odyssey, just one with a less explicit grandeur of vision than *2001*. *Gravity* takes a tinier microcosm as its "*mise-en-scène*." But the essential structure is the same. Both might bring us back: to our senses; to this ball in space; to our place in the cosmos, *without*, one might hope, the same dangerous grandiosity that we started with or developed.

To conclude: What good are the spacecraft shown drifting magnificently through space in these films, unless one can return? In its early part, *2001: A Space Odyssey* seems to offer a rationale for technology, with the first appearance of the monolith coinciding with the first invention of a weapon (a weapon that flickers seamlessly into the shape of a spacecraft, in a famous and revealing moment of Kubrick magic). But what happens with HAL implies perhaps that a sufficiently complex technology—of the kind we are now deeply reliant on—will eventually fail us. As appears to be happening, today, as techno-fixes fail to stop our self-destructive trajectory.[52]

It is important (then) that *2001* ends with its direct "address" to viewers, a second-person I-thou moment. Just as *Gravity* ends with an alienation effect (the wonderful mud shot of the Earth hitting the camera), so, more patently, does *2001*. The Star Child's gaze out of the fourth wall, perhaps recognizing us, and our living planet, demands a response from us.[53] But, one might ask, what kind of response? Well, "at minimum" (!) it seems to be asking us to awaken. It asks what it would actually mean (and require) *for us* to be reborn. That is: to truly *come back* to Earth; to walk the Earth fully aware of her beauty and fragility, *as if* we loved her and our fellow beings enough to (have made the effort to) return across a whole universe to do so.[54] To place hyper-technology back in Pandora's box, to stop racing towards a deadly future and to be fully present, awake and appreciative in the NOW.

I think we *know* already the answer to the question *2001* in effect ends by asking. The problem is more whether we have the imagination and the *will* to act adequately, after, first, responding with mature thought and authentic feeling.[55]

Notes

1. Hannah Arendt, The Human Condition (Chicago: Chicago U. Press, 2013 (1958)), pp. 1–2. (Emphases added.)
2. Unlike Robinson Crusoe and his Hollywood descendants, including the character played by Tom Hanks in *Castaway*, and the one played by Robert Redford in *All Is Lost*.
3. Note that the space shuttle has a woman in charge—like everyone else in the shuttle, though, she gets killed in the first impact of debris.
4. She says, in a moving (and self-pitying) speech to the radio ham who can't understand English: "I'm gonna die; . . . I know, we're all gonna die. . . . But I'm gonna die today. . . . Nobody will mourn for me." (I examine this speech at greater length later in this chapter.)
5. See Charles Taylor (1985), "Atomism," in Philosophical Papers (Cambridge: Cambridge University Press), pp. 187–210. See also my www.opendemocracy.net/transformation/rupert-read/how-whales-and-dolphins-can-teach-us-to-be-less-stupid.
6. All of this is reminiscent of the Ray Bradbury story "No Particular Night or Morning" (in The illustrated man (New York: Doubleday, 1951). A man suffers terrible loss on Earth and goes into space to disconnect himself from everything that could produce further pain, eventually denying the very existence of the past and of ever more aspects of the present; in the end he drifts into empty space in his space suit, accepting only the existence of his own mind. The difference is that Bradbury's story is a meditation on scepticism as to other minds (or solipsism) as a disastrous philosophical challenge, whereas *Gravity* is interested in quasi-solipsism as an (un-)ethical, self-protective temptation. It's the difference between something that can be lived only at the cost of psychosis and something that can be lived more easily—at the cost of neurosis. Stanley Cavell (in The Claim of Reason) famously describes this as the difference between madness and tragedy. *Gravity* is interested in the latter. For, in this respect, Stone

is close to Justine, more than to Kelvin; close to Elle, much more than to the haunted/haunting narrator of *LYiM*.

7. Not coincidentally, I think, her last words to him (to the person she remembers) concern her daughter; she asks him to look out for her in the afterlife. She mentions her daughter's missing red shoe to him, and that she has it. The notion of a pair of red shoes is resonant, and much could be done with the resonance. I will restrict myself to noting that in what might be claimed to be the most totemic of all "hero's journey" movies, the great *The Wizard of Oz*, it is the (again female) protagonist's red shoes that enable her to magic a connection back to home. And what is *Gravity* showing us, if not that "There's no place like home/Earth. . . ."
8. I go into the kind of point in play here much more deeply in Chapter 6, in relation to *Lord of the Rings*.
9. Obviously, this might be compared with the same issue in *NLMG*, especially at the very end of that film, as discussed in Chapter 3.
10. Stone's daughter is dead, and grieved for. But Stone herself achieves a kind of rebirth in the course of the film. And my argument below includes that we, potentially, do too. Thus in a sense Stone's heroine's journey in the film can be seen as a kind of (gender-neutral) pregnancy. (As indeed Nietzsche frequently wrote of himself as pregnant.)
11. See the argument of my "Does grief have a logic?," in O. Kuusela et al. (eds.), Wittgenstein and Phenomenology (London: Routledge, 2018).
12. I.e. what I mean by logic is not a system of symbols, but simply the way that we think and speak when we are doing so adequately, i.e. an adequate way to think about our concepts. This is similar to Wittgenstein's notion of "grammar," or the practical constraints and actual nature of language. (See my and Phil Hutchinson's "Grammar," in Anat Matar [ed.], Understanding Wittgenstein, Understanding Modernism [London: Bloomsbury, 2017].)
13. In this way, the reminder is like the kind of reminder that Wittgenstein highlights philosophy as presenting us with, at Philosophical Investigations 127. One isn't reminded, in philosophy, of ordinary facts: one is reminded of the very—unnoticed, normally—basis of one's existence. Cf. also PI 79: "Say what you choose, so long as it does not prevent you from seeing the facts. (And when you see them there is a good deal that you will not say.)" "The facts" here should not be understood as referencing things such as facts about the chemistry of oxygen, or about the battle of Austerlitz, etc. .
14. This short film "Aningaaq" is worth watching: http://mashable.com/2013/11/21/gravity-aningaaq-short-film-oscars/. This short is also included on the *Gravity* DVD.
15. Kundera, The Unbearable Lightness of Being (New York: Harper, 1999 (1984)), p. 5.
16. I borrow "ownmost death" from Heidegger, who uses the term "ownmost" to emphasize that death is non-relational, it is mine and mine only (cf. Being and Time (Oxford: Blackwell, 1962) 50: 294). The awareness of my own death as the possibility of my not-being, constantly present, defines me as a finite being, and discloses my self authentically.
17. This film is an antidote to the disastrous, anti-ecological, anti-gravity fantasy of Nolan's *Interstellar*: a film which could quite literally be called "Anti-Gravity." See Peter Kramer's "*Interstellar*: An initial response", http://thinkingfilmcollective.blogspot.co.uk/2014/11/an-initial-response-to-interstellar.html for more on what is so wrong in *Interstellar*. While *Gravity* pulls us back to our home, *Interstellar* falls precisely into the trap picked out by Arendt in my opening quote: it tacitly treats each successive planet as a temporary source of resources, to be used up and moved on from.

18. We might see *Gravity* as posing a new equation: rather than, as Simone Weil supposed, taking gravity and grace to be obviously and implacably opposed, we might *equate* them. We might say: Gravity, properly understood, *is* grace.
19. Of course, it is also possible to read the contrast between angels and ourselves as a basis for praising the fact that we are earthly; our "finitude" opens us towards the other, and makes essential the close relationality that is one of my central topics in the present work. (Thanks to Mihai Ometita for reminding me of this.)

    A masterful reworking of this theme of how to understand the contrast between ourselves and angels is to be found in Philip Pullman's His Dark Materials (which is *inter alia* a reworking of Paradise Lost).

    Note too that Ryan Stone, as she "talks" to the ghost of Kowalski, explicitly calls her daughter, Sarah, her "angel"; and Stone is here talking of Sarah as (if she were) alive, as in the present tense. People do of course not infrequently describe those dearest to them as "my angel."
20. As in the long "therapy" scene in the bar in *HMA*, on which I commented in Chapter 2.
21. This moment will strike some viewers as corn. That is, to some extent, how it has often struck me. But I think, in its defence, we should note that this film, like *Avatar* (see Chapter 6), is addressed to ordinary Americans, taking a reference point that will widely appeal to average viewers—like the appeal of going on a "wild ride"—and reinterpreting it as a metaphor for life itself. Thus *Gravity* can return us to life. As Stone is returned to life, and even her dead daughter and dead Kowalski are, in her. The elite sneerings at *Gravity* and *Avatar* as simply mass entertainment, the assumption that a 3-D film cannot be serious—cannot be art—are themselves defences against the way that these films, with remarkable boldness and ingenuousness, offer us a possible path to freedom from precisely these assumptions. And, more broadly, a path towards freedom from the pernicious aspects of American politics and anti-ecology that these films critique. Stone is not a philosopher and nor should she be. She is Everyman (Everywoman). You, midwifed by the film, are the philosopher.
22. The view of the Earth as a whole, which first became possible only 60 or so years ago, was a new view into its finitude and vulnerability. Whereas, as one might put it, exploitation is driven by a horizontal gaze: by the assumption that there's always more to exploit—behind those hills, across the sea, over that (ever-receding) frontier. Roughly: if the Earth is flat, then maybe we don't have to worry so very much about ecology and the limits to growth; but on a round Earth, we do.
23. A documentary film called *Collision Point: The Race to Clean Up Space*, concerning this pollution, appears on the DVD of *Gravity*. The film features various leading scientists and technologists, as well as Cuarón, and is narrated by Ed Harris. This space-pollution is gently pre-figured very early in the film, where Stone, "forgetting" that she is in space and that objects there do not behave as they do on Earth, accidentally lets a bolt float away. It almost becomes a piece of space debris—until Kowalski stops it from doing so, by retrieving it for her.

    An intriguing point about the scenario examined in *Collision Point* and dramatized in *Gravity* is that, if it did occur, it would pretty much permanently put paid to fantasies of space exploration/colonization etc.—because it would become desperately hazardous for us ever to venture out in space again. We would have surrounded ourselves with a blanket of ultra-high-speed space junk that would be virtually permanent. A fitting way to end the space age, perhaps: enforced Earth-dwelling for the species chronically unable to feel settled here or to provide the Earth with good stewardship.
24. It is known as "the Kessler Syndrome": a chain reaction, exponential growth (of debris) causing more and more destruction. (Does this remind you of anything? Consider human economic growth.)

25. Noting of course that Stone earlier almost dies, as Kowalski will die, of an excess build-up of CO_2 inside her helmet. The parallel with human-triggered climate change is clear. (Compare also the atmosphere of Pandora in *Avatar*, as discussed in Chapter 6.)
26. Once more, echoes of *2001* here. See the closing portion of this chapter.
27. See Ben Child's "Gravity 'celebrates presence of God' say U.S. Christian reviewers", www.theguardian.com/film/2013/oct/10/gravity-god-george-clooney.
28. Which she looks at, with a wonderful arbitrariness (given that she is in space), earlier when "talking to" Matt up in Heaven about her daughter.
29. In fact, "alienations" are present, I would suggest, at every key moment in the film, every turning point in it. At the moment analyzed earlier of the reality of spinning away towards death in space, the "long take" passes impossibly through the barrier of Stone's visor; a different kind of "aesthetic suspense"—of suspense about whether the filmmakers can carry off something which seems blatantly unrealistic—suffuses the scene of Kowalski's absurdly improbable return; and then the film ends with this remarkable mud moment.
30. Thus the crucial moment at the end of the film with the muddy water hitting the camera as she triumphantly emerges from the water is an "alienation" effect in that it most starkly reminds you, the viewer, of your point of view (cf. also n.29, above).
31. Consider the final dialogue between Stone and Kowalski, as he floats off "voluntarily" to his death:

    KOWALSKI: So now that we have some distance between us, you're attracted to me, right?
    RYAN: What?
    KOWALSKI: Well, people say I have beautiful blue eyes.
    RYAN: [Pause] You have beautiful blue eyes.
    KOWALSKI: I have brown eyes . . .

    . . .
    KOWALSKI: Oh, my God . . .
    RYAN: What?
    KOWALSKI: Wow. Hey, Ryan?
    RYAN: Yeah?
    KOWALSKI: You should see the sun on the Ganges. It's amazing.

    "Gravitational" attraction; attention, being (in) the moment; it's all here.
32. Think for instance of her asking the radio ham to make his dogs bark again for her. Even when she is about to die, and inclined to give up on life, there is something left to enjoy.
33. Arendt, The Human Condition, p. 2.
34. Of course, *Gravity* is not really a "sci-fi" film in the conventional sense of that word. It is not even "tech-fi." There isn't any fictional science relevant to the plot in *Gravity*, nor even fictional technology. *Gravity* is simply a story that occurs mainly outside our atmosphere, and that ends back on Earth.
35. And with this move, of course, we circle back to *Hiroshima, Mon Amour*.
36. And here is how *Gravity* and *Hiroshima* teach something about the failure to learn, which is the fate of the protagonists in *Marienbad*, and arguably also of *Solaris*, whose ending similarly illustrates, as argued in Chapter 4, above, what it is to be psychotically trapped. *Melancholia*, *Hiroshima* and *Gravity* concern neurosis: depression, through unresolved grief. But happily that is usually resolvable, and in each case we see the film offering a path to resolution. (For a very thoughtful account of *Melancholia* side by side with *Gravity*, see Christopher Peterson's "The Gravity of Melancholia," *Theory and Event* (2015) 18:2.)
37. In his Reasons and Persons (Oxford: Oxford University Press, 1984), Part III. For a full account of this, see my "The tale Parfit tells," *Philosophy and Literature* 39:1 (2015).

38. Peter Kramer suggests that, in Arthur C. Clarke's novel, "[HAL] has to keep the true goal of the mission secret from the two astronauts [and so] it suffers a mental 'breakdown'." (2001: A space odyssey, London: BFI, 2010, p. 14). In the script that Kubrick and Clarke worked on together, it was the secret that led to HAL's breakdown. As Kramer writes, "Viewers might conclude that HAL . . . is troubled by the fact that 'he' cannot share [knowledge of the Moon monolith and its signal beamed to the Jupiter/Saturn area] with the astronauts. . . . As a consequence, he seems to start malfunctioning" (p. 71). However, the lack of explicitness on this point in the film is important: whatever is not in the text, one ought not to project into it with definitiveness.
39. Including Kramer.
40. Compare here the way that (in Chapter 6) I suggest we might start to be able to empathize with the somewhat similarly near-all-powerful and malign Sauron in his fear, in *The Lord of the Rings*.
41. The humans in the film have some emotions (not many), but only for each other / for themselves. A clear counter-example to this might appear to be Bowman's furious repeated "Open the door!" command to HAL, as he seeks to regain the spaceship. But I think that Bowman's frustration here is really, ultimately, with himself—for repeating the command when in his heart he knows that HAL will not obey and cannot be influenced by emotional blackmail. Another possible counter-example might be Bowman inviting HAL to sing him a song, at the end. But here, again, we can interpret his actions merely as an attempt to shut HAL up and keep him occupied, so that he can shut him down ("kill" him).
42. One of the wonderful things about *2001* is the amount of real time that passes in silence. When one looks at or listens to the dialogue, much of it is banal (has virtually no meaning). It is/offers the kind of boilerplate that we exchange with one another in trivial/"professional" public social situations but that does not really carry meaning. The humans in the film, as a whole, lack personality: the only real personalities drawn, one might claim, are "Moon-Watcher" the ape-man-innovator, and HAL the computer.
43. The "Star Child" of course impossibly conquers these long distances, these unsurvivable non-existent supply lines, at the film's end. The Star Child *symbolizes* perhaps the need for possible return from any journey. Our need for an emotional and inter-personal rebirth, but also, on the largest scale: our deep need for a resilient healthy Earth as a reliable home. The desperately long "supply lines" of *2001* (and *Avatar*) ought, in other words, to make us reflect on the excessively long supply lines that already characterize an (i.e. our) excessively economically "globalized" world. On this, see also the close of the present chapter.
44. Kramer's 2001 helpfully points out the various ways (including the malfunction of the antenna linking them back to Earth) in which the astronauts—and, I would add, HAL—are so horribly isolated, are suffering such an extreme of unprecedented loneliness (Kramer, 2010, pp. 68 and 82).
45. Cambridge, MA: MIT, 1992.
46. This thought is present in the work of Merleau-Ponty and Wittgenstein, more explicitly even than of Heidegger, I would argue. See also Graham Button et al., Computers Minds and Conduct (Cambridge: Polity, 1995).
47. Though it is not clear that the postulation of artificial memories, as in the case of Rachel, actually makes sense. It is not clear that one could come to have genuine sophisticated mentation on the basis of fake "memory implants." Not clear that there is any substitute for genuine growth/learning.
48. I owe this observation to Tom Greaves.
49. As I have suggested above, the greater level of explicitness in Clarke's 2001 book is (on balance) a grave weakness. The film, forcing one to encounter a

mystery and to figure out for oneself what could be going on, as I have tried to do here, is more transformative, more (in Wittgenstein's terms) "therapeutic." The film puts one in something like the position of those who encounter the monolith; the book at times gives one something rather closer to a God's-eye view, which is not, as I see it, what the story is supposed to be about/for.

50. *2001* is an odyssey, we might say, through time and emotional maturation, just as much as it is a journey through space. This becomes explicit in the long takes at the very end of the film (and this yields yet another connect between *2001* and *Gravity*: their profound dependence on long tracking shots). Cf. on this point also n.53.
51. The Star Child returns perhaps to deliver peace, starting with the end of our nuclear arsenals. That, at any rate, is what appears to happen at the end of Clarke's novel. (See n.54.)
52. The following are powerful readings on this point (which is not to say that I agree with everything in them; I don't): Clive Hamilton, Earthmasters: The Dawn of the Age of Climate Engineering (New Haven: Yale University Press, 2013); Kevin Anderson's "The hidden agenda: How veiled techno-utopias shore up the Paris Agreement", https://kevinanderson.info/blog/the-hidden-agenda-how-veiled-techno-utopias-shore-up-the-paris-agreement/; Tim Radford's "Stop burning fossil fuels now", www.theguardian.com/environment/2015/aug/03/stop-burning-fossil-fuels-now-no-co2-technofix-climate-change-oceans and www.washingtonpost.com/wp-srv/national/longterm/unabomber/manifesto.text.htm?tid=a_inl_manual.
53. This shot echoes the very many shots where we stare directly into Bowman's face, in the course of the film. But somehow those shots never seem quite as extraordinary and meaningful as the Star Child's gaze addressing us. For the Star Child is awakened. He is modelling what we need to undergo.

    We should also note here the beautifully slow "relay race" of points of view that happens in the final several minutes of the film, after the "Star Gate," an extraordinary representation of long slews of time in just a few minutes of screen time. And perhaps too the fascinating fact, discussed above, that there are point-of-view shots from HAL's point of view in the film.
54. In the ending of Clarke's book, which is of course rather different from that of the film, it seems evident that the Star Child, the neo-Bowman, has some kind of protective intention towards the Earth, and in particular that he wants to keep the atmosphere "cleaner" (London: Arrow, 1968), p. 256. As with *Avatar* (see Chapter 6), the real question here is of course what we will do to try to realize the idea of the art, if that idea impresses us, given that we can expect no such *deus ex machina*. (I.e. no interventionist deus—and no machina either—to do the job for us.)
55. Thanks to the ThinkingFilm collective, and (especially) to Jerry Goodenough and Naomi Marghaleet, for conversations and interactions that have improved this chapter. Thanks to Mihai Ometita for comments, some of which I have raided to improve my thinking on *Gravity*. Huge thanks especially to Peter Kramer, for co-authoring with me the blog post (http://thinkingfilmcollective.blogspot.co.uk/2014/01/gravitys-pull.html) which gave birth to this chapter, and for his great work on *2001*, and for comments.

6 The Fantasy of Absolute Safety Through Absolute Power

The Lord of the Rings Trilogy (and *Avatar*)

> Suffering begins to dissolve when we can question the belief or the hope that there's anywhere to hide.
>
> —Pema Chodron

> The desire for safety stands against every great and noble enterprise.
>
> —Tacitus

> The venture is great. Only a deep and mighty faith, permeating a person's whole being, is equal to it. It is a faith of a unique kind, different from trust between man [*sic*] and man, a faith which reaches out to the whole of things and can do no other than stake all it has.
>
> —Nicolai Hartmann, in "Love of the Remote," on love of future people

What explains the enduring and quite vast appeal of *The Lord of the Rings*? How and why is this book—more recently made into three fabulously successful and (in my view) deeply impressive films—able to touch parts that other epics cannot reach?[1]

In what follows, I will combine film analysis with a good deal of contextual material to do with madness, morality, politics and history, in order to generate a perhaps surprising answer to this question. I will lay out reasons to believe that *The Lord of the Rings* carries off a profound exploration of a psychological, political and philosophical issue of almost incalculable importance. I suspect that most of Tolkien's readers and (still more so) most viewers of Peter Jackson's film trilogy implicitly sense this, and that the vast appeal of this neo-myth is thereby explained better than it has otherwise been. The "reading" that I will essay here of *The Lord of the Rings*, this allegorization of my own readings/viewings of it, is I think a reading that captures and makes explicit a major dimension of this great story that may in part at least gradually come to strike most readers and viewers as indeed having been already an important part of their experience of the work, even if they had not exactly realized so prior to reading my essay.

As with Chapter 4, this time it's personal. The film trilogy came out during a time in my life when I was in slow recovery from depression, including experiences of derealization (the reality of the world, or of aspects of it—such as its beauty—seemingly withdrawing), depersonalization (temporary unworlding loss of sense of self), paranoia, panic and so forth. It struck an enormous chord with me from the very beginning, and a whole new aspect was opened up to me which I had never noticed in the books. Which in turn shed light for me on philosophy of mental health (which had already been a main research and teaching specialization of mine). So the interpretation given below draws on film-viewing, on familiarity with the relevant academic literatures, *and* on first-person life experience.

Moreover, it turns out that the personal is political, as well as psychological (and philosophical). As I shall explain. But first, I need to spend some time laying out the fundaments of *Lord of the Rings* as I found it, psychopathologically speaking.

The briefest way to try to sum up my reading of *The Lord of the Rings* [*LOTR*] is as *an exploratory allegory of serious mental suffering* generally (and of madness in particular), the suffering that comes principally from aversion to one's own mental states. Especially, an allegory of the mutually reinforcing character of paranoia and of the withdrawn and "split" state of "schizoid" consciousness, mimicked or attended and further mutually reinforced by (mutually reinforcing) anxiety and depression. I believe that *The Lord of the Rings* book and (more so) the films ought to be viewed as an investigation of these, and particularly of the way in which more or less schizoid withdrawal, while promising to enhance one's safety levels, actually tends to fuel one's fears until one reaches prodigious levels of anxious dread.

It may seem absurd to credit this "children's story" with such an ambition. All I ask is that you give this interpretation a try. I am going to claim that (especially) Jackson's films[2] actually do succeed astoundingly well in deepening our understanding of the phenomenology and the mechanisms of serious "mental illness," *precisely because* they are not in the slightest didactic. They are *applicable* to their topic (the word "applicable" being the word that Tolkien himself, in the Foreword to the second edition of *Lord of the Rings*, much preferred to "allegorical"), rather than too overtly or explicitly schooling one in it. They work as metaphor, as "allegory," in the way I am urging, precisely because they lead one to *experience* vicariously the motivations for and dilemmas of madness—"the paradoxes of delusion"—[3] without insisting that they are doing so. Much as the person who finds themselves descending into madness does not *know* that that is what is happening to them; although they usually think that it perhaps is (their efforts to sort out what is really happening to them typically becomes in itself an important part of their condition, their problem, sometimes *producing* a delusional system). In other words, in a positive riff on *Catch-22*: if you are quite certain that what is happening to you is simply a mental disturbance, then it need not worry you that much. The deeper terror of the person going mad is that they are not going mad, but that *this* (what their experience seemingly points towards, what they find themselves starting to believe etc.) *is really happening*. Ergo, a phenomenologically effective imagining one's way into madness must not be too *knowing*. A film explicitly and plainly *about* madness could never effectively capture the experience. (And this entails that, if my way of taking *LOTR* is right, it *has* to be the case that it isn't *obvious* that it's right.) Taking up an external position on madness[4]—as a film explicitly about madness necessarily does—*necessarily* fails. Because in madness one lacks precisely such an external perspective. (And this too is why a film like Bergman's *Persona* has the peculiar form it does.)

And we experience this, in *LOTR*. Right from the sense of strangeness—the sudden need to scrutinize and to hide—which constitutes a rising tide of perplexing open-ended anxiety, of schizy trouble, a kind of "praecox" disturbingness, as soon as the Ring makes its presence felt at the start of the story, all the way to the tragic departure of the Ring-bearers from the consensual everyday world, at the end.

Wittgenstein held (see e.g. *PI* 133) that philosophy (of the right kind) is required in order to alleviate the kind of mental suffering that philosophy (of the wrong kind) produces. Being tormented by questions that are based in confusions and in runnings-away from reality.

Thus my way of taking ("applying") *Lord of the Rings* indeed deserves to be understood as a philosophical one.

The process that leads one to put on the Ring, is, I shall suggest, a speeded-up version of the descent into psychosis that results *inter alia* from the kind of vicious internal retreat—the kind of splitting and re-splitting—that R. D. Laing famously characterized as the main mechanism, through schizoid withdrawal, of schizophrenia.[5]

Or—and it is deeply interesting that this comes to much the same thing—a real-time version of the vicious and borderline-psychotic circle into profound feelings/intimations of unreality ("derealization") that can very rapidly result from extreme cases of anxiety-overwhelm, of panic. It is not so much that Frodo puts on the Ring when he is *in* a panic, though that is indeed true; it is more that the entire process of wanting to put on the Ring and of finding that after a while it doesn't help—in fact, the reverse—*is* panic. Panic squared, to the point even of psychosis.

How do the films, on my account, generate this kind of effect?

Consider the very opening of the first film, *The Fellowship of the Ring*. "The *world* has changed," a narratorial voice tells us. This is exactly what the sufferer from an incipient madness is inclined to think. That what has changed is not me, but *the world*.

We see and feel all this in part because in due course the power of cinema shows us this very different world. In particular, a world which is *ex hypothesi* available only to the Ring-wearer is shown to us, and by no means exclusively through point-of-view shots. We see Frodo in this world, a world only visible to him. (I shall return shortly to the importance of this paradoxical cinematic point, and how one is to understand it.)

And in that "world," of course, one almost "meets" the Lord of the Rings.

Who/what is the Lord of the Rings?

As understood psychoanalytically, Sauron is the return of the repressed in spades. He is the vengeful father who will condemn the relatively tiny and powerless you *for what you are*, in the innermost core of your being. The further you retreat, the less of you there is and the more of Him.

We might add that He could be your conscience, alienated from you by yourself and set over against yourself to condemn you. For if you condemn yourself, you are at least inoculated against disappointment in the sense that you cannot fail in any difficult quest or task worse than you have already condemned yourself for failing.

He is the ultimate nightmare of madness, the never-so-rational fear that God might turn out to be malevolent. What if there was an all-powerful all-seeing being who was consumed with malice towards you? There might be one; so you had better invent Him now, before He takes you unawares. (If God had not already been invented, He would nevertheless have been invented. He has probably been invented over and over again in different cultures at different times; for the God of dread is a natural result of subjecting psyches to severe strain.)

Or as understood roughly after the fashion of the innovative post-psychoanalytic psychologist, Louis Sass: Sauron's ferocious, merciless gaze, that unstill Eye with no eyelashes or eyebrows,[6] is the product of excess thinking, *not* of a turn away from Reason. The product of (self-defeating) attempts to find safety and power through giving up on ordinary safety/trust; through withdrawal into the self. (As will emerge more fully below: I mean both that Sauron could become like this himself through such a process, and that he could be invented by others through such a process.) Such that, to quote one of Sass's patients, "In my world I am omnipotent, in yours I practice diplomacy."[7] The withdrawal leads to a loss of reality, and a self-perpetuating and multiplying sense of unreality. If one withdraws far enough, one starts to wonder whether one exists at all, and either then hypothesizes that the parts of one that are "not really me" are an alien consciousness, or hypothesizes that there must be a creative—and probably aggressive, would-be sovereign—Other who is bringing one's consciousness into existence or taking one's consciousness as an object.[8] If one then finds the voices one hears and the eyes one feels searching one to be real, that's hardly surprising, and in part happens because they seem at *least* as real as ordinary people etc. in the consensual world seem—that is, perhaps, not very. In the world of the Ring-wearer, everything somewhat flickers, and is not quite tangible.

According to Sass, the mechanisms of deep neurosis and (in some ways still more so!) of psychosis are "rational" mechanisms. The world of the "mad" is mostly not a world of Dionysian abandon, nor of primitivity or regression (as many psychoanalysts would have it), but nor is it a simple product of cognitive deficit and defect (as many contemporary cognitive psychologists and psychiatrists would have it). It is a cold, dead, hyper-scrutinized[9] "world" where all that is alive are powers of death and morbidity and observedness. It is a world of thinking and observing and fearing without check. Alone in this "world," it is all too easy to dream up the worst of possible companions, or to speculate that one's own all-too-tiny, puny and fragmentary finite self can hardly be the foundation of all these experiences—maybe they are being given to one by a highly powerful and malign demon. In this "world," as one retreats to what is comforting or certain and tries to blot out the rest, it is all too natural to try to scrutinize oneself, too, to catch signs of infection, invasion, illness, or badness—and for this scrutiny endlessly then to try to catch its own tail. As one feels watched by an alienated part of oneself, naturally, one keeps trying to retreat further—unless one manages in time to take off the Ring, and to return perhaps to the ordinary fears and perils and reliefs and pleasures (and banalities) of (ordinary) life.[10]

How exactly does all this get shown on film? For, as admitted above, it is not as if we see all this from Frodo's point of view, literally.

The first answer, before entering the "world" of the Ring, to set the mood as it were, is that we hear the Ring dimly almost speaking to us. We almost-hear the not-quite-words of its imagined power. Furthermore, in a brilliant touch, Jackson gives us several point-of-view shots from the point of view of the Ring. I shall return to these points below.

The fuller answer is, I think, this: the most effective way usually of showing the mind on film is *not to show what the mind in question is perceiving, but to show the face, to show the body*. Wittgenstein helps us to understand this, in remarks of his such as "The best picture of the human soul is the human body" (*PI* "Part II" section iv) and "The face is the soul of the body" (*C&V* p. 23e).

We *naturally* tend to empathize/identify with the person—the mind, the soul—shown us on-screen, at least if the film (and the acting) is good enough. (And Elijah Wood does a fabulous job of giving us Frodo's lovely and tortured—familiar—mind. *Our* mind.)[11] We *naturally* enter into Frodo's world, which is the mental world we learn most about in *Lord of the Rings*—even (in fact, especially) when we are being *shown* <u>Frodo</u>, rather than <u>what</u> Frodo sees and experiences.

This is an important pointer towards what is I think a terribly important (Wittgensteinian, counter-Cartesian) philosophical point: we are NOT typically *stuck* in our heads, isolated definitively from one another. When we ARE, that is *in itself* pathological.

In the key scene at Weathertop in *The Fellowship of the Ring*, when five of the Ring-wraiths almost win the Ring from Frodo and almost turn him into a wraith himself, we see events (once Frodo has put on the Ring) mostly from a third-person vantage point, but as if we are wearing the *One* Ring, which is of course quite impossible on the text's own terms. We see what it is like to be *in the world of the Ring-wearer*. This is the way the language of film works. And—because?—this is the way that human beings (normally) work. They/we can identify with others, easily, if we let ourselves, not through literally seeing things from their point of view, but simply from genuinely seeing *them*.

It is important that the overwhelming odds in the fight scene on Weathertop give one a sense of "hopeless" dread. The four pitifully prepared and tiny hobbits are rapidly encircled by the five huge, demonic, powerfully armed and utterly ruthless—completely focused—Ring-wraiths. There is no chance for the former. It is important that one feels this—one is then feeling what they start to feel, and *what precipitates the dash to some desperate psychic safety*. Frodo's donning of the Ring shows us and gives us a sense of a kind of instant and very serious (and quite reasonable) traumatic-stress disorder.

This effect is only multiplied as the films go on; again and again, its protagonists are tempted to give up hope, and sometimes express sheer despair, even if they do not usually *practice* that avowed despair. For example: the scene on Weathertop in *The Fellowship of the Ring* partly anticipates the scene at Helm's Deep in *The Two Towers* where the awesome and merciless Uruk-Hai army—now, a whole *army* of incipient death and complete destruction—stands arrayed before Helm's Deep, surely destined (as they almost do) to wipe out this race of Men. That scene in turn anticipates the scene at Minas Tirith in *The Return of the King* wherein the incalculably vast army of Mordor threatens to overwhelm inexorably the remnants of Gondor. The Orc leader surveys Minas Tirith and remarks, with satisfaction: "Fear. The city is rank with it." At the wonderfully ironic order, "Release the prisoners!," his soldiers begin the terrifying assault by releasing the decapitated heads of Faramir's cavalry troops as cannonballs onto the city. Things go from worse to worser (to worst?) as the Ring-wraiths arrive once more, this time on their ghastly flying steeds, screaming in a way that invades the mind, dealing death and terror left and right, especially horrible in their tactic of picking up men and hurling them to their deaths now as human cannonballs, with those "cannonballs" thus taking others within the city with them as they die. These scenes, I am suggesting, are successively ratcheted up[12] evocations of dread, of a fear that starts to free-float into an almost hopeless terror that can then dream only of escape; most powerfully and attractively, of an inward escape. The kind of escape enacted by Denethor, Steward of Gondor, in the relatively (!) "ordinary" fashion of depressive and at times semi-psychotic denial of the facts, and by Frodo either through putting on the Ring or (as especially at Osgiliath, as we shall see below) through the tempting yet potentially yet-more-disastrous expedient of simply giving the Ring to the enemy. Actually, Denethor's method is in a key respect structurally the same as Frodo's. Denethor resists disappointment *by moving fairly directly to the outer reaches of disappointment*. Here, the film investigates and displays the mechanisms and attraction of depression: as we saw already in Chapter 4, if you make yourself sad first, then you are aiming to be less vulnerable, and this is more true the sadder you make yourself. Denethor assumes that Faramir is a failure. He assumes later that Faramir is dead, despite evidence of life in the latter. (Or possibly, in a twisted kind of caring,[13] he is beating the enemy to the chance of achieving total defeat through killing Faramir, by killing him first.) He assumes that "Rohan has deserted us"—despite it having been *him* who has strongly resisted even calling on Rohan to help. This is a deeply pitiable condition to be in; Denethor too, in fact, demands our sympathy and even love. (I shall return to the full meaning of these expediencies—and to the full meaning of the retreats within the structures of Helm's Deep and Minas Tirith ordered by their anxious rulers—below.)

We see *as* Frodo, then, when we are shown Frodo in the grip of the Ring, in, as one might put it, "Ring-world." And we understand that you put the Ring on for safety—and what you find is a deeper terror, a world of obsession and terrifying loneliness, a world where, alone with yourself, you are mercilessly watched by an Other, and thus a world where it's harder to breathe even than it was in the terrifying situation you have just "left behind."

We get a very strong sense of what it is like, too, as Frodo's journey nears its end. As, in climbing Mount Doom to rid himself of the Ring, he is so under its power that he barely exists any more. (It is almost as if he is now already wearing the Ring, even before he gives in and grandiosely claims it for himself and actually does put it on, inside the volcano.) Sam tries to remind him of the Shire, to bring him back. Sam asks Frodo if he can remember the taste of strawberries. Frodo's rasping and increasingly terrified reply runs thus: "No, Sam. I can't remember the taste of fruit, nor the sound of water, nor the touch of grass. [I'm] Naked. . . . There's nothing. No veil between me and the wheel of fire.[14] I can see Him . . . with my waking eye!"[15] Not to be with others, but to be alone with yourself, with your fantasies, *is* to be with another (but another with terrifying powers, not a friend), *is* to be dangerously divided. From—and yet tantalizingly, horribly close to—what you desire above all and fear above all.

As the proverb has it, power corrupts, and absolute power absolutely. As David Loy puts it, "Tolkien's Frodo cannot use the Ring because it would use him."[16] Or, in Simone Weil's terms:

> Power contains a sort of fatality which weighs as pitilessly on those who command as on those who obey: nay, more, it is in so far as it enslaves the former that, through their agency, it presses down upon the latter.[17]

But we might hazard, given what I've argued so far, that *The Lord of the Rings*, superficial appearances notwithstanding, is not in the *first* instance about (political) power and its corrupting effects. It is, perhaps, first, about psychopathology. It would be then about the former largely only *through* being about the latter. It is about the pathological effects of the desire for absolute safety, which are found macroscopically in the political realm but which have roots in—and can easier be dissected microscopically in—the psychological realm. The desire for absolute safety is found in a pure form in various "mental illnesses." It is found in a *structurally* identical form in the desire for (total) personal power in politics, and in the desire of states and rulers for (total) power over other states or peoples.

The Lord of the Rings is, I would suggest, about kings and prime ministers and presidents and rivals and subjects and terrorists *through* being about "mental illness" and the pull to withdraw from the consensual world, the real world.

Let us go further into how *Lord of the Rings* is about madness: through Jackson's fine-toothed investigation of the character of Gollum. Gollum embodies—or at least, seems to (we shall return to this point)—the loneliness and corruption yielded by the obsessive and addictive lust for "power." Gollum never wanted to rule any kingdom; but he wanted the greatest Ring of Power for himself, and was prepared to go to very violent lengths to get it.

Jackson's films provide a marvellous portrayal of the split mind that results from a no-holds-barred push for such power/completion, such possession of what is most "precious." What *is* most precious? What that you could seemingly have and control could be more precious than *security* for you and yours? Love might be more precious; but love cannot be guaranteed. It depends on another. But can't one's own security at least be guaranteed? What *Lord of the Rings* suggests is that, if one goes down this path, one will reify one's desire for security into the ultimate security blanket, the One Ring, which Gollum loves and even feels loved by, and which he of course treats neither as an aspect of himself nor as simply a piece of metal but *as if it were a person*, as if it had a voice and a personality and intentions and could literally be loved, and communed with. And that one will *not* then feel secure—Gollum is "for instance" deeply haunted by Sauron's gaze.

We see Gollum's quasi-multiple-personality vividly portrayed in his private dialogues with himself. *And* we see him, like Sauron,[18] feeling incomplete without the Ring. The second of these is more classically schizoid, more like what we discussed earlier with reference to what one's experience is like once one puts on the Ring. The first, dissociative kind of "split" ("dual personality"), we should now integrate into that discussion.[19]

The Two Towers, the second film in the trilogy, fundamentally centres upon *Frodo not giving up on Gollum*, and *Gollum not giving up on himself*.[20] Why is this so important? Because Frodo has to believe that even Gollum could "come back." That he (too) could recover. (This is perhaps why Frodo says, "This creature is bound to me, and I to him.")

There is a historical parallel worth exploring here at a little length.[21] It is with various ways of "touching bottom" that are profoundly feared by humans, as perhaps has been most vividly visible in the cruel laboratory of the concentration and extermination camps. Take the following remark of Primo Levi's:

> [W]e, the survivors, are not the true witnesses. This is an uncomfortable notion, of which I have become conscious little by little, reading

> the memoirs of others and mine at a distance of years. We survivors are not only an exiguous but also [an] anomalous minority: we are those who by their prevarications or abilities or good luck did not touch bottom. Those who did so, those who saw the Gorgon, have not returned to tell about it or have returned mute, but they are the "Muslims," the submerged, the complete witnesses, those whose deposition would have a general significance. They are the rule, we are the exception.[22]

How can and should we further imagine the "Muslim" or "muselmann" from the camps, those so named because of the posture they frequently took up, bowed "completely" into submission by the horrendous processes that they were subject to?

- Possibly as one reduced to what might be termed an unconscious moral nihilism, a desperate, completely self-centred existence, where there is nothing left to life except bare survival?
- But most often, in the testimonies of Levi and others, we see the muselmann not as a human reduced as it were to such a social Darwinism, but rather to something without much even of that last strong spark of life. A lassitude descends; even the *spirit* of bare life has gone from the "person."

Both these possibilities—being *in the grip of* one's own efforts at prolongation of bare life, efforts so desperate that there is no remaining capacity for or interest in reflection or communication etc.;[23] and life that continues but without any efforts any more at its own prolongation, (instead) *mere* continued existence—can be imagined as unspeaking, as varieties of human being reduced to mere individual biological life. Both possibilities, obviously, are reasonable objects of dread to the person contemplating them, and contemplating becoming thus.

The dread of oneself becoming either of these "stages" is, I submit, necessarily in part a dread of becoming *stuck* there, or becoming stuck on some descending trajectory. The fear is that touching bottom will lead with inevitability to becoming *stuck at* bottom. If one could recover, then it wouldn't be quite so bad; but the thought of not recovering at all is the thought of losing oneself completely, to disintegration, death, madness or evil. (And these perhaps no longer seem so different from one another.)

Take this epochal moment, from *If This Is a Man*. Levi has just heard the last of the heroic crematorium-destroying Auschwitz rebels crying, "I am the last one!" before being publicly hung. He feels shame at having done nothing, and feels that this is the last *man*, who has just been killed:

> Alberto and I went back to the hut, and we could not look each other in the face. That man must have been tough, he must have been made of another metal than us if this condition of ours, which has broken us,

> could not bend him. // Because we also are broken, conquered: even if we know how to adapt ourselves, even if we have finally learnt how to find our food and to resist the fatigue and cold, even if we return home.[24]

Levi fears that he has become the moral nihilist, or perhaps the quasi-solipsistic Darwinian near-"Muslim." He does not in his dreadful current circumstances have the support perhaps necessary to see that *the very existence of his fears gives the lie to their content.*[25] The fact that he cares about and fears his current state (of inaction, of "failing" to act in a "dignified" manner etc.) proves that he has *not* been utterly broken.

It is very worth noting that the point here is internally related to the point discussed earlier about the deep unsatisfactoriness of films etc. which in a knowing way explore (or rather exploit) madness from the outside, and on films *necessarily* not seeming to be (only or even primarily) about madness if they are to count as successful explorations of madness "from the inside." The difficulty then, a difficulty wrestled with increasingly in Levi's work, is how one can come up with a *criterion* for what counts as an effective and accurate account of or depiction of the extreme experience of the camps—or likewise, I would add, of madness—without falsifying it. For it seems that an account of such existential conditions from the inside is in principle impossible. An accurate account must be from rock bottom. *If* there can be any account at all from such a place—which would seem, to say the least, unlikely, if madness is the complete absence of a work, if being a "muselmann" means precisely not being able to give an account, because one has left the world of striving and commonality—then, at the least, such an "account" will not seem to us to be an account of what *we* take the condition to be *at all.* An accurate depiction of madness will not seem to be about madness—and thus its accuracy is hard to judge (for there are many, many films that perhaps do not seem to be about madness at all!). An accurate depiction of the muselmann will not seem to be about the "muselmann" at all—the same difficulty follows. It seems then that at best we cannot *identify* a true witness to these things. (Levi's thought, quoted earlier, that the drowned would be the only true witnesses, now appears additionally supported.)

I explore and aim to resolve this serious difficulty, so far as madness is concerned.[26] The line of thought I am tentatively pursuing in the present chapter places me in a critical relation even to Levi. I am committed to arguing, absurd as it may sound, that in a certain sense there is/was no muselmann, no once and for all moral nihilist, no human reduced successfully to something decisively subhuman. (Being killed is not being so reduced.)

For any would-be muselmann could not be completely at ease with their state. But to really be the muselmann we postulated, to be drowned while alive, one would have to be non-resistive to that state. Therefore there is no such thing as the muselmann sunk definitively in the would-be muselmann state.

The figures of the lost madman, the psychopath, the pure addict, the muselmann, terrify us. But the terror is, in the end, of our own making, and, *so long as we can be terrified in this way, there remains an element of us which is clearly unconquered.* It makes no sense to think of humans as bare biological life with a rational/mental overlay, which can be stripped away to leave the biological animal substrate bare once more. Such a complete splitting of mind and body, of rational and animal, is a crucial part *of the very pathologies and disasters which it claims to diagnose.*

My line of thought is a radicalized extension of the line of thought central to Terrence Des Pres's magnificent and neglected book, *The Survivor*, a book that is among the great works of secondary literature on the concentration and extermination camps. Des Pres suggests that sociality and ordinary virtue *are a part* of us, indissolubly. Arguably, even Hobbes may have believed this too;[27] but our fear is such that, rather following the likes of Freud, we tend to read Hobbes as saying that the natural condition of human life is "nasty, brutish and short," when what Hobbes was perhaps actually saying was that this was the condition of human life only *when such life is reduced to chaos*, and humans become atomized from one another. But Hobbes set the scene for liberal political theory, which has (regrettably) dominated Anglophone political philosophy almost ever since, with his atomistic metaphysical *picture* of human beings, unless they are unified under a sovereign, their relation to whom is primarily one of fear. (Strong echoes or presentiments of Sauron here.) As in Hobbes's version of religion, which is of religion stripped of spirit. Sauron is nothing if not a material god (or a material demon). Tolkien's world is a "Hobbesian" world, with the sovereign/God become a material analogue of the Cartesian malign demon (or at best, become a real-world post-*LOTR* Aragorn—see the close of my text, below, for more on this). This "sovereign" is an ultimate object of fear; a being, who you ideally want to become rather than be subject to, who is material (and thus real), malevolent, overwhelming and yet dependent upon you; you are not *entirely* abject. (So: *you can still fail.*)

We tend to fear that, when the "veneer" of civilization is stripped away, then *Homo sapiens* will be revealed as an ugly individualist. The camps were apparently an experiment that showed that this was true—or so at least many have thought. But, while the camps were in many regards *designed* to try to get their denizens to behave in this way, Des Pres and others have shown that many still did *not*, and indeed that many who seemingly thought that behaving selfishly was the only way to survive in the camps were proven to be quite wrong. Compare the following remark, from a Treblinka survivor quoted in Gita Sereny's *Into That Darkness*: "[I]t wasn't ruthlessness that enabled an individual to survive—it was an intangible quality . . . a faith in life."[28]

"Faith in life" is one very appropriate label for my rendition of what the antidote to the temptations wrestled with in *Lord of the Rings* is. Des

Pres goes on to remark, with reference to the virtual disappearance of sex, of erotic drives etc., in the harsher camps, "If this runs counter to Freud's view—that civilized rather than primitive conditions repress erotic need—so be it. Behaviour which does not support day-to-day existence tends to vanish in extremity. We may fairly conclude that what remains"—in which category he included the virtuous action, the sharing of food for instance, which in some camps was virtually an *index* of the relative likelihood of the person's survival—"is indispensable."[29] And here is where I would extend or radicalize Des Pres's proposal a little further: Wrong though Freud is to think of civilization as primarily a represser of supposed natural near-overwhelming selfish tendencies, it is inaccurate to try to disprove his thoughts about civilization in repressing erotic drives in the way Des Pres seeks to—for what one needs to stay clear about is that the circumstances in the camps were by and large, surely, much *worse* than "primitive" circumstances. Our pre-civilized living conditions were surely very, very rarely, if ever, as bad (all things considered) to live in as the hells that totalitarianism has created. The conditions in places like Treblinka and Birkenau and Auschwitz were hells only made possible through a high degree of quite deliberate "rational" thought and planning. What remained "indispensable" to people within them is *less than* the bare minimum we should ascribe to life, even to "primitive" life. Such life surely—certainly—includes sexual life, as well as what is arguably the still more basic form of life that Des Pres (and others) rightly highlight: communal life, organized activity, ordinary social virtue: "For animals as for man, return to community is an inborn reaction to danger and prolonged stress. *Only under highly favourable conditions can a society tolerate anti-social forms of behaviour*. We can pretend we owe nothing to anyone, but survivors know they need each other."[30] (This rebuke to liberal and libertarian political philosophy is one that I see *The Lord of the Rings* as embroidering on.)

But to return to Primo Levi's line of thought, above, which is, psychologically speaking, a pretty compelling one: if something is infinitely important, as one's own existence or sanity or decency can easily, naturally, come to seem to be, then one will be engrossedly concerned in and with ultimate threats to it. And one will risk self-fulfilling engulfment by such threats if one becomes hopelessly caught up in one's engrossment.

On the surface, this dynamic is explored in *Lord of the Rings* primarily in moral terms: Can one come back from losing one's centre, one's conscience (as Gollum is presumed to have done, in his possession by the Ring)? But, I submit, this is structurally no different from the very same dynamic explored in psychopathological terms, in terms of the loss of one's sanity. I shall pursue both lines of thought together here, as I think that in fact both are—*simultaneously*—at play in the story.

Why must Frodo believe that Gollum can come back from a state of insanity, or psychopathy? Well, Gollum was once like him. Ergo, he could become like Gollum.[31] He feels the weight of the Ring, feels its pull getting harder and harder to resist. His rational fear helps him to reach the conclusion: I *must* believe that I could come back even from the worst imaginable excesses of madness or loss of self/conscience that this very hard situation that I am in might induce. Because if I think I could not, then this will be an object *worthy* of my infinite fear, of dread. And, given that I am already feeling anxious and unstable, it will be impossible not to focus fearfully on this fear: I will then be launched on a self-reinforcing journey to *total terror*, to madness. (This negative self-fulfilling prophecy is again what the Ring yields, for one who enters deeply into its field of force.)

You may have nothing to fear but fear itself—but if it feeds on itself to the point of disabling dread, as it very easily can, then that's *quite* terrifying enough, just by itself, to lose you everything. (At least: that's what the fear "implies" or whispers—or shouts—to you, when it is upon you.)[32] Fear of madness and plain dread feed on each other, and threaten rapid implosion to insanity, unless one holds onto believing that even insanity is not necessarily interminable. (Hope feeds on hope, too; and faith on faith.)[33]

So Frodo indeed must believe that he will be able to come back. And thus he must believe that Gollum can come back. And thus he allows himself to feel pity and even love for Gollum. In part, to save himself. Frodo fears he is becoming Gollum. Gollum represents—he does not *instantiate*, but he psychically *represents* or symbolizes—the human becoming in- and/or subhuman. He is—to the eyes of humans who shy away from him, who withdraw from him, who leave him alone to fear and hate (for instance) his own self, in a self-fuelling circuit of distrust and alienation—the definition of a monster. As Levi feared he was becoming the very kind of hateful thing that the Nazis said he was. Less than a worm, a selfish and unworthy subhuman, one who had forfeited ties with real humans.

And my philosophical claim, then, is that in the sense in which Levi and Frodo and all of us fear touching bottom, our fear is groundless—except, ironically, as potentially self-fulfilling. *There is no compelling reason to believe* that anyone cannot come back from the temptation to moral nihilism, from profound selfishness, even from a desperate or desolate withdrawal from life altogether. It is human to be appalled and terrified by the thought of becoming subhuman; *as long as one has the capacity to have such a thought, one has not so become.* Change their circumstances, and we have no reason to believe that anyone is eternally incapable of such reemergence. Not even the profoundly oppressed, not even those whose will to life has been sorely sapped; and not even one's worst enemies.

This I think symbolically explains also an apparent plot flaw, an instance of silly "unrealisticness," in *The Two Towers* (perhaps especially in the film version). Namely, the release of Wormtongue after Gandalf's dis-enchantment of Theoden. Wormtongue should have been killed or at least locked up, according to the warrior ethic of so much of *Lord of the Rings*. It was obvious that he would most likely go back to Saruman and cause further murderous mischief. So he does: many of the men of Rohan die when he then cleverly urges Saruman to set the "wolves" of Isengard on them, as they are *en route* to Helm's Deep.[34]

The irrational pity of Aragorn towards Wormtongue mirrors or in fact exceeds Bilbo's pity towards Gollum (stressed to Frodo by Gandalf when they first see Gollum, in Moria), when he had the chance to kill him long before. The *point* of such pity is that *one ought not to give up completely on any human being*, or indeed on any hobbit or their ilk.[35] (I discuss below the way in which Frodo's pity—and care; in fact, love—for Gollum is in the end all for the best.)

You can perhaps give up on the enemy races and species—and here is the danger of racism endemic to *Lord of the Rings*. (I shall try to remove the sting from that accusation later, in part by suggesting that even they should not be given up on.)[36] But a key "message" of the trilogy is that at minimum you mustn't give up on anyone recognizably or even just potentially human. And that you must keep on working to carry out such recognition: that that is an ongoing *task* or journey of *acknowledgement*.

It is of some interest that it is *Aragorn*, who I want to suggest is a kind of counterpart or alter ego for Frodo throughout Jackson's trilogy, who orders the release of Wormtongue. I have suggested that *The Two Towers* fundamentally concerns *Frodo*, and his pity for and fellow-feeling with Gollum. How then can *The Two Towers* possibly work as a film? For it consists of three entirely distinct plotlines. The protagonists of these three plotlines never meet on-screen, and indeed end up further from each other than when they started. This sounds like a quite disastrously incoherent text.

And yet the film does work. Some viewers, myself included, feel it even to be (on balance) the best of the three. Why and how is this?

The issue central to *The Two Towers* is found in its most focused and condensed form in the Frodo-Gollum nexus, but it is also very present in the plotline involving Aragorn, Gandalf, Wormtongue and Theoden. Compare Gandalf not giving up on Theoden, despite the latter's deeply sunken aged unreachability, his being possessed by Saruman as others are possessed by Sauron, his being deeply withdrawn to a place of "safety" where he can be secure in his kingly "power."[37] (Perhaps Gandalf can only do this because he [Gandalf] let go [at the chasm, in Moria], because he "died" and yet didn't give up.)[38]

What of the third narrative, of Merry and Pippin? The same applies. How?

Aragorn refuses to give them up for lost, but that structures his narrative, not theirs; for they quite "outrun" Aragorn's efforts, and are eventually saved by Treebeard, and they then recommence their struggle, with the latter on their side.

Merry and Pippin are, as Treebeard puts it, "very small," and, though becoming brave, are less committed to the quest than Frodo. They are genuinely tempted to give up and retreat to the Shire as per Treebeard's suggestion after the Ents refuse to help them, refusing (at first) to recognize themselves as (as Merry puts it) "part of this world."

But Merry realizes that *they can't retreat*; that the disaster will eventually come and find them there, too.

The deft inter-cutting from scenes featuring Gandalf and Aragorn, to scenes featuring Merry and Pippin, to scenes featuring Frodo, in *The Two Towers*—cutting not present in the book, which is just really a collection of two or three different narratives stuck more or less sequentially together—points up clearly, I think, what I am speaking of. This is *one* film. The protagonists may be physically far apart. But their story is one: they are as one in wrestling with how to keep faith *in themselves*, in others, in anything and everything.

***

The positive way forward in the psychological struggle that is at the heart of *Lord of the Rings*, and perhaps especially of *The Two Towers*, finds a *macroscopic expression* in Merry's rousing of the Ents' ecological conscience. The manoeuvre of retreating to the Shire would have been *the same as* the manoeuvre symbolized by and shown in the Ring, in the putting on of the Ring.

And such retreat would be a giving up of the wholeness that is present in ecological consciousness, in an understanding of one's groundedness in the Earth and one's fellow-ness with one's fellows (and ultimately, I will suggest, with all creatures). The Ents staying put in their paradise, the hobbits retreating to theirs—these would be merely fools' paradises, fallacies/fantasies of safe havens. The fear of political and ecological catastrophe causes Pippin to retreat, much as the Ring encourages retreat; but a deeper and more challenging and truer survival instinct kicks in in Merry, and, after a little cunning on his part, the struggle to save Fangorn and the Shire and the world is on.

And it is the same again with the retreat to Helm's Deep. This time, not ecological wholeness but the construction of the false self and its overcoming (to achieve a psychical wholeness, and a unity with others) is foregrounded. What do I mean? Gandalf calls on Theoden, recovered from his withdrawal, to ride out and meet the enemy in the open, on the plains. Theoden will not. So recently having ridden out of the recesses of his mind to meet the world again, his first instinct is to retreat once more. They recess to Helm's Deep, to a fortress deep inside a ravine, with no exit. Aragorn calls on Theoden at least to send out riders, to seek aid from elsewhere. Theoden will not. He battens down the hatches. Marvellously, some solidaristic aid comes anyway (Elrond's Elves, who are roused twice in the story by Galadriel's caring—her recognition of all of our being-with-others—to offer their aid to "the world of men"), though probably not enough.

And there, in the form of the battle, we see a very vivid metaphor for the retreat and yet further retreat which is the lot most strikingly of schizoid psychotics (but not only of them). The very structure of Helm's Deep, an ultimate defensive fortress, is a visual presentation of a mind seeking to find somewhere where it can reign supreme and not have to confront monsters face to face. The inexorable retreat inwards does not help, however. It only prolongs the inevitable, and perhaps makes it worse.

How does the tide of the battle turn? When, utterly improbably, the riders of Rohan do ride *out* from the very keep of the castle, at last taking the attack to the enemy, out into the open of a dawn.

The three "separate" stories of *The Two Towers* are one. The skilful editing work in Jackson's film shows this, to those who have eyes to see (it).

Whose task is the hardest? Merry and Pippin's, who pass through mortal fear at the hands of Uruk and Orc (and Fangorn) before going bravely to battle, in a seemingly doomed cause, with the Ents, against Saruman's very base?[39]

Or, more impossibly difficult still, perhaps: Aragorn's, in not giving up on those hobbits, and then Gandalf's and Aragorn's, each of whom apparently *die* before going out (not in) to fight against inconceivable odds, against dread, against Saruman's vast army?

Frodo's. Ordinary life, companionship and the building of trust (including, crucially, in oneself), achieved not through the more extraordinary version of these that is ideologically involved in being a warrior, is the hardest of all.[40] The ordinary semi-private task of not giving up where the not-giving-up in the face of great temptation is a daily—almost continuous—occurrence. And where one is deliberately going out—journeying—to face voluntarily one's greatest terror.

Not fighting, but *giving up* one's weapon, and in the course of so doing, coming to know oneself in all one's possible conditions, is the hardest task of all. The Merry-Pippin-Treebeard and the Gandalf-Aragorn-Theoden plotlines are, in the end, roughly, "objects of comparison" for the Frodo-Sam-Gollum narrative. The first two conclude successfully by the end of *The Two Towers*. The last continues, harrowingly and remorselessly, on into the third film, and with a ghastly unnamed threat now hanging over Frodo and Sam,[41] through Gollum's loss of faith in them.

Why does Frodo's task keep getting harder? By the end of *The Two Towers* Gollum has more or less fatally lost trust in him. But of course that isn't all of the problem, or even most of it. We must look a little closer at the question of why in the first place the power of the Ring gets stronger the closer one gets to Mordor.

A wonderful text to read alongside *The Lord of the Rings*, in this regard, is Franz Kafka's magnificent and surprisingly little-known novella, "The Burrow."[42] The protagonist and narrator of this story is a creature or human (it is unclear which) who has designed for himself a huge burrow, in which he hopes to be completely safe. Surprise, surprise, his feelings of insecurity are not banished by the burrow's construction. So much so that from time to time [. . .]

> I almost screw myself to the point of deciding to emigrate to distant parts and take up my old comfortless life again, which had no security

> whatever, but was one indiscriminate succession of perils, yet in consequence prevented me from perceiving and fearing particular perils, *as I am constantly reminded by comparing my secure burrow with ordinary life.*[43]

He desires complete safety, the kind of guarantee sought after by a Descartes. If only, he reflects, he had someone else he could trust, who could watch over the entrance to the burrow while he himself went down and took refuge inside it; *then* he could *feel* secure:

> If I only had someone I could trust to keep watch at my post of observation; then of course I could descend in perfect peace of mind. I should make an agreement with this trusty confederate of mine that he would keep a careful note of the state of things during my descent and for quite a long time afterwards, and if he saw any sign of danger knock on the moss covering, and if he saw nothing do nothing. With that a clean sweep would be made of all my fears, no residue would be left, or at most my confidant. For would he not demand some counter-service from me; would he not at least want to see the burrow? . . . But I simply could not admit him, for either I must let him go in first by himself, which is simply unimaginable, or we must both descend at the same time, in which case the advantage I am supposed to derive from him, that of being kept watch over, would be lost. And what trust can I really put in him? Can I trust one whom I have had under my eyes just as fully when I can't see him, and the moss covering separates us? . . . No, if one takes it by and large, I have no right to complain that *I am alone and have nobody that I can trust. . . .* I can only trust myself and my burrow.[44]

Like the Ring-bearer, the burrower is by (self-)definition, alone. And the latter then starts to dream "my dream of a completely perfect burrow, and that somewhat calms me."[45] Only "somewhat," of course, for *there can be no such thing as a "completely perfect burrow"*; for it would have to be a burrow that ensures total security. But insecurity is a restless beast; it is not a matter of designing something; it is a state of mind. It is, to the burrower, a constant and growing torment.

Sometimes, the burrower of Kakfa's "The Burrow" thinks that it must be through some moral fault of his, some "sin," that he fails to find peace. And indeed, we should probably emphasise more strongly than we have so far, by way of psychologically interpreting *The Lord of the Rings*, the truly huge role of guilt and shame in much depression and anxiety—and, *sometimes devastatingly magnified*, in psychosis. Deep feelings of worthlessness, of oneself being some kind of abomination, of one's deservedness of being othered or punished or annihilated: Gollum perhaps enables us to understand these better than we did before.

There is a natural vicious circle between such terrible guilt, or shame, and the profound fear of a malevolent demi-God (Sauron) and the desire for a safe refuge ("Ring-world," as I've sometimes termed it—that "perfect burrow" available to one instantly at the movement of one's fingers). Not only are such overpowering guilt-feelings etc. a natural *cause* of such dread, in ways that we have already made perspicuous (e.g. What is the worst thing that could happen to someone who has done something terrible/who is something terrible?), they are also a natural *consequence* of such dread (e.g. How bad must I be, if "God" is prepared to devote so much effort to getting *me*?). Gollum's self-loathing, his aloneness and alienation, his internalization of the negative image others have of him (note the way the words "Gollum! Gollum!" are sometimes almost vomited out of his mouth, involuntarily, as an unwanted self-description/self-denigration, blocking access to a happier past-self that Frodo tries to revivify): all these and more point up how Gollum is always at risk, through a moral-psychological process, of losing himself, even of losing his mind. And so, by extension, is Frodo.

If you give up on any human being, you give up on humanity. If you give up on humanity, you might as well not bother fighting or trying anymore; you have already lost what you hoped to be defending. And in any case, if you give up on any part of humanity, you implicitly give up on yourself. As Donne might have put it, if he had read Tolkien: ask not for whom the Ring calls or shines: it is for you. Frodo mustn't give up on Gollum, or he may give up on himself, and the *fear* then is one of being *lost* in psychosis. As suggested above, this is a moral point on or near the surface of *The Two Towers*; I am suggesting that at a deeper level it is a psychological point, a deep observation about psychopathology, and about the possibility (and difficulty) of recovering therefrom. (A severe difficulty in recovering fully from psychosis is, I believe, the very great difficulty one who has so lost their footing as to have a psychotic episode has, understandably, in convincing themselves that they could cope if they had a worse one. Or even the difficulty in feeling confident that one has in reality recovered at all from the earlier episode. This is the true terror-horror of "flashbacks": the thought/feeling that one never really emerged at all from the traumatic situation one thought was past. "Flashback" is in fact therefore rather a misnomer, from an experiential or phenomenological point of view; "flashbacks" are not terrifying through being a vivid reliving of a past experience; that wouldn't be so very bad. Rather, one feels as if the experience in question *is not past at all*, and that it is unterrifying ordinary life which is a fantasy, or wholly in the past. The deep terror of "flashback" is *not* that one is flashing back; it is once again that *this* is really happening, and that it is *ordinary life* that is the delusion.)

When one feels close to losing all faith, in others and in oneself, then one is most vulnerable to feeling under threat from a powerful external agency. A *loss* of faith naturally leads[46] to the fear of . . . not quite God. Rather, the devil as God.

Or again: not quite Satan, but Sauron? In the godless and virtually religionless world of *Lord of the Rings*, Sauron is of course the closest we get to g/God. An overwhelmingly powerful and malevolent external force. A malign genie who has gradually seemed to incarnate, to take a quasi-physical form. Just such a malign being is one of the most important things that one fears/finds/*encounters*,[47] at peak times on "stimmung" or psychotic journeys.[48] The desperate search for safety *results* in one being overpowered by an overwhelming dread at an overwhelming watching, judging, heartless and destructive external agency. The search for safety *results* in one seemingly being confronted by absolute nemesis, with no expectation of being saved by a benevolent force—there is none as strong, or none that is willing, one is convinced. That God is onto me, and that "God" is a malevolent demon—just that super-Cartesian possibility is, I am urging, lived out at the deep dark heart of *Lord of the Rings*.[49]

In fact, *Lord of the Rings*'s analysis is far subtler and more psychologically real than Descartes's emotionless academic rendition of the mind "meditating" upon the terrors of possible cosmic aloneness and the company one might surprisingly and regrettably find oneself keeping in that aloneness.[50] For Tolkien/Jackson, the God-awful malign demon is not a self-standing ontic thing. Rather, to be God-powerful *it needs something to complete it*.[51] It needs, I submit, *you*, or more specifically, your fear and desire and weakness (that we witness corroding everyone in the film, and sometimes, as with Saruman, succeeding).[52] It needs your desire for power, that corrupts, that takes you from others; it needs your self-fulfilling fear of "it"; it needs your weakness, that would hand the power over to "it" in a doomed bid to lessen the grip upon you of dread. The malicious demon (in Tolkien) *depends on you*. He is not *all*-powerful, without the One Ring. You are not nothing beside Him;[53] you are just pitifully small and vulnerable in comparison, as you toss on the sea of fate. He will only *become* all-powerful if you try to become him, or alternatively simply give him the power he seeks.

The rationale here, and it is a profound one, is this: if God/Satan/Sauron quite simply is all-powerful, then—and paradoxically—your worries are significantly tempered. For there is then no quest, no chance of escape, nowhere to hide. One can give up worrying.[54] The mind in search of absolute safety, the mind unused to not-worrying and unwilling to *risk* such a state, must then restlessly pass on *beyond* the assumption of one's absolute abjection before God.

If one were infinitely less than God, then one would be to some extent relieved, even if God is malign or is some kind of "demiurge":[55] *because at least there is then nothing more one can do*. One can simply sit back and wait to be annihilated or tortured etc., safe in the knowledge that there is no way out. Belief in an omnipotent God, if the God may have a worse than Old Testament temper, is a means to the psychological security of not actually having to go on actively worrying and acting. The mind in search of absolute safety, the mind in search of any possible threats to it will quickly, restlessly, move on: the more worrying thought that comes to one next is that perhaps one does still have a part to play, that one's actions will be consequential, that what one does or thinks next could make things even worse. Paradoxically, there is something even worse than abjection before an all-powerful malevolent demon: namely, the threat of a less than all-powerful malevolent demon whose power and action *depends on you*, on what you do and think. The ceaseless hungry terrified motion of schizy thought is here: Tolkien correctly identifies a potentially self-fulfilling threat to thought and to one's very sense of identity more profound than—and a logical *extension* of—that which Descartes set out for us. (This then is literature/film *as* philosophy: Jackson and Tolkien have taken us somewhere new, somewhere undreamt of in Cartesian philosophy.)

And it is *the very same dynamic* that is depicted in the scenes of overwhelming odds, of dread in the face of malevolent power that threatens your obliteration, that we described earlier: Weathertop, and the awesome battles of Helm's Deep and Minas Tirith / Pelennor Fields. These scenes, paradoxically, are in a key sense *worse* than facing an *omnipotent* malevolent agent would be: for *you*, including your failings, still have a role to play in them; you still hope, and thus still properly fear (a great deal); you have yet to be completely disappointed. What is happening in these scenes is a visceral improved version of the thought experiment of the malevolent demon, replete with the full terror of such a demon not wanting just to play elaborate party tricks on you but to enslave and torture (or at best, to murder) you and all your kin, and furthermore of it very much seeming like "This is really happening" (but with a lingering uncertainty about whether it is or not, an uncertainty that is in itself often described in analogous cases by schizophrenics as literally tormenting), *and* of the demonic powers being almost completely overwhelming, but of their leaving you a slim frightening chance of saving yourself—frightening, in that you still have the chance to fail, and thus to be the one to blame for losing all for humanity.

Am I playing fast and loose by seeming to move seamlessly here from (1) the thought experiment that Descartes invites his readers to participate in, to meditate on and in, to (2) the dire thoughts of "schizophrenics" and other painfully suffering mentally ill folks, and (3) the scenarios depicted as real in *Lord of the Rings* of battles in or after which oneself and all hope are overwhelmingly likely to suffer torture and/or obliteration? I believe not: because I believe that psychological torture—e.g. what is suffered by some "schizophrenics," the agony of psychical dying without rebirth, or of imagined/"remembered" torture—can be just as bad for the sufferer as real torture. This might seem disrespectful of or ignorant of real torture. But I suggest that, on the contrary, the suggestion that my belief on this point is false is likely to be predicated upon ignorance of just how fantastically appalling purely mental/"imagined" torture can be. And bear in mind: the worst aspect of actual physical torture is often said by sufferers to be the anticipation of its continuance; it is that that is most responsible for much "post-traumatic stress disorder" among recovering victims of torture. PTSD in such cases is I think an utterly understandable watchfulness for the possible resumption *or sheer continuance* of the torture (again: "flashbacks" would be essentially trivial if they *presented* as daymares; what they actually are experiences that the sufferer, unworldingly, takes or fears to be *present reality*); and it is a (self-defeating) hyper-guarding by the individual against the overwhelming terror and disappointment that such continuance would bring. (This phenomenon is strikingly similar to the guarding against the possibility of a malign demon etc. that a rational person who is subject to schizoid withdrawal—or something akin to it—will subject themselves to.) *Lord of the Rings* shows us the ultimate price of such eternal—hyper—vigilance: paranoia, madness. PTSD is being in such a state of alert against feared present/future traumatic assaults upon one that one cannot appreciate that one is now actually safe.

As Terrence Des Pres writes: "[W]hat we experience symbolically, in spirit only, survivors must go through in spirit *and* in body. In extremity, states of mind become objective, metaphors tend to actualize, the word becomes flesh."[56] The state of mind of one suffering from the grip of psychotic doubt and delusion can be for them very concrete, very "objective." The death-camp inhabitant suffers from the word becoming flesh, from actually experiencing the nightmares and hells that we imagine. And often, this experience does not leave them once they have left the camp. The psychotic suffers from an over-spiritualization, from the flesh becoming word—but this too can yield an *experience* which is similarly quite as if it *were* the nightmares and hells that the rest of us at most *imagine*. The one process is the obverse of the other—but that doesn't mean that they are not *both* equally *torturous*.

Indeed, I would submit that the only difference between hell and heaven is a state of what we might call "confidence" on the part of the human subject. Confidence in oneself (though not what often passes for "self-confidence" in our culture, a kind of egoic brazenness and closure from others), confidence in others (that there is "some good in the world," as Sam memorably puts it plain), confidence in life. Con-fidence. With-fidelity. What I am calling here true faith. Without which one can feel utterly lost. Heaven and hell are states of mind, with a fine, fine line between them, and all the difference in the world resulting. For a state of mind can be the purest torture.

Compare Des Pres's account of a survivor of a mass execution:

> Bullets did not tear through her, her heart did not stop. But she was certain—her body was certain—that death was coming. She felt that she had died, she lay for hours among the lifeless mass of her comrades, and then got up. Is this the famous valley of death through which souls pass? Is this resurrection? How much is metaphor, how much plain fact? Or is there any longer a difference? // . . . Man's interior drama, the height and depth of spiritual experience, has been writ large in the world. The concentration camps have done what art always does: they have brought

> us face to face with archetypes, they have invested body with mind and mind with body, they have given visible embodiment to man's spiritual universe.[57]

Yes; only psychosis *had already done this*. Some human beings had already suffered infinitely variegated forms of torture before those forms of torture were actually invented, and even, I am sure, similarly, death camps before the latter were dreamt of by mere politicians and generals and Eichmannian bureaucrats. Des Pres says that the survivors "felt themselves die and then return to life," and the "objective correlative" of their ordeal was not a symbolic representation or a ritual entered into imaginatively. It was the world itself, albeit a world such as *we* know through art and dream only.[58] And through the experiences of those who do not merely dream such a world for others, but, involuntarily, for themselves.

Is poetry, art, the novel, dead, after Auschwitz? And/or: Have I left Tolkien and Jackson behind, now? Hardly; my point has of course been that, quite unexpectedly, their art helps those of us, a mass audience—who have not experienced psychosis (nor its partial anticipations in the phenomenology of neurosis: such as derealization and paranoia), nor the torture of the mind that can seem a torture of the body etc.—to gain some greater purchase on it, in its hellish strangeness, its tormenting unreality, its awe-ful presentiment of what is happening or is about to happen. Their art works with and through these same archetypes that Des Pres mentions; it dreams for us a confrontation with the reality and the unreality of such horrors as survivors and psychotics alike have experienced. The deepest reason why this perhaps-implausible-sounding parallel is justified is, I claim, the following: that Sauron and all his stand-ins (e.g. the Nazgul) are or at least might as well be *the creations of the mind of Frodo* (*and the creations of the mind of the viewer* of *Lord of the Rings* just insofar as one can enter imaginatively into the mind of someone passing through a major psychotic episode [which, judging by the film's success, seems to be: perhaps surprisingly far]).

For *what it's like* is I think stunningly visually captured in (for instance) the scene late on in *The Two Towers* where Frodo, in the battlements of Osgiliath, is confronted by the Lord of the Nazgul on his fell winged beast. With the Ring-wraiths near, Frodo withdraws. He goes into a state of quasi-schizoid withdrawal, leaving the consensual world half-behind as the power of the Ring is upon him. He starts to enter a private and almost silent "world." He is tempted to put the Ring on; he stumbles towards the Nazgul lord; we witness and are ourselves breath-stoppingly scared or awed by the silent confrontation of Frodo by this vast foe. Being before and "held by" such an (apparently external) foe is just what being on or just over the cusp of a psychotic break can feel like; Jackson evokes this marvellously with the eerie stillness that descends for a few moments (as again, in *The Return of the King*, before the Lord of the Nazgul's city, Minas Morgul) in this situation of the knowing relation between the emissary of malevolence and his

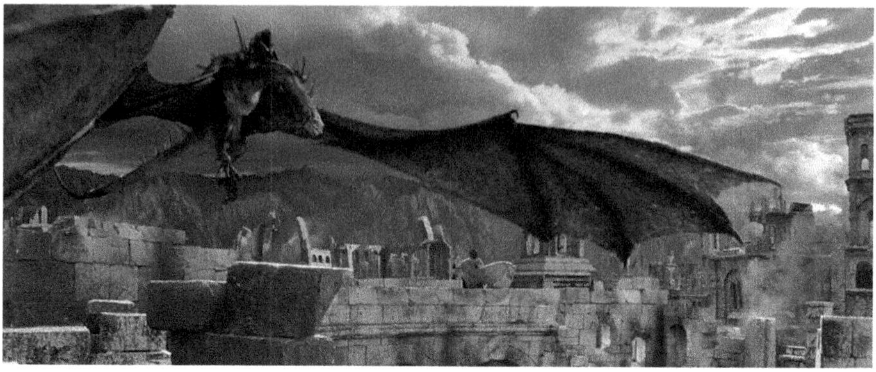

Figure 6.1 Facing a vastly (but not infinitely) greater foe: Frodo and the Nazgul.

prey. Frodo then wants to lose the burden of the Ring, tired of and desperate to escape from the entire situation in which he has found himself, and makes as if to give it to the king of the Ring-wraiths. Feeling undone by the overwhelming hostile external force that he does not realize he has himself called to him, he seeks for a moment to be rid of it that way, too.

Saved on that occasion (from the compelling oscillation between the two compelling—though in reality equally ineffective—alternatives of putting on the Ring or giving it away) by Faramir and Sam, Frodo is allowed to journey on, the appalling nature of his burden now well-understood by Faramir, towards Mordor. Faramir agrees to let Frodo and his companions go on, despite its being against the law to do so, when he *hears* about how the Ring drove his brother Boromir mad (Frodo's words), and finally when he *sees* this extraordinary image: Frodo with the Nazgul looming vast and terrible over him. He *sees* the way the Ring is too strong even for this determined (tormented) little man to bear.[59] The image of the Lord of the Nazgul on his appalling winged steed towering over Frodo is more terrifying than any image of a bluntly all-powerful demon could be. Because it is an image of gross disproportion, an image of the relentless massively powerful heartless hunter set over and against the small and tormented and deeply vulnerable hunted; and yet because, as things then stand, at Osgiliath, *they can still get worse*.[60] If things cannot get any worse, then at least one is immune to disappointment. (Whereas if one's moral or psychological infirmity—one's decisions—will complete something entirely appalling, will bring into being something more monstrous than yet exists, then . . .)

Faramir thus chooses to let Frodo to go on—to let the Ring go—rather than to seize it as a weapon as Boromir had sought to do. Faramir lets Frodo go on, to dissolve the weapon and its creator away, rather than trying to fight the enemy, "the Dark Lord," with his own weapon. Faramir's earlier speech before the body of a dead foe, just after he first captured Frodo and company perhaps indexes directly, if paradoxically,[61] the point about the non-existence, except as constructed by us, of the enemy who looms over us in our minds, who we fear and that we then (and self-defeatingly) seek power over: "The enemy? . . . I wonder who he was, where he came from, and if he was really evil at heart."[62] We should note the audacious placing of this remark of Faramir's: It directly follows Frodo saying to him, "Those that claim to oppose the enemy would do well not to hinder us." I.e. Faramir here anticipates my ultimate claim about *LOTR*: that there *is* no enemy; that evil has no positive existence; that our enemy is only the traits and temptations that *lead* to war (and that we must never forget the presence of these in us, too).

Faramir marvellously (because counter-hegemonically) refuses to give up on this man who fought against them, and he asks whether what they are fighting is anything more than mirror images of themselves.[63] Other beings, seeking safety through violent power, just as we (are inclined to) do. If they exist at all, if they are not paranoid creations of ourselves, then they are like us anyway. The only really existing enemy is a being like us, who should not, by rights, be our enemy. We come to understand the men fighting against "the goodies," we come to understand the brutalized Orcs and Uruks, we come to understand the wraiths, and Gollum, *we come to understand even Sauron*, through Frodo's (and Faramir's, and Gandalf's) powers of empathy—and through one's *need* to not be alone. But when the power of the Ring is upon you, then even if these "enemies" do not exist, in your aloneness you will soon powerfully invent them. When you put on the Ring, and become invisible, you become visible to your darkest fears—the wraiths,[64] and God/Satan/Sauron.

Faramir and Frodo and Gandalf are saying in effect that what allows what appears to be evil to "manifest" most strongly in or as a being that faces one, is precisely the self-delusive belief that Evil simply *is*. Actually the scarier thought, *and* the saner thought (though it can precipitate a deeper psychical crisis), is that bad things happening or otherwise is in part my/our ongoing *responsibility*. That evil doesn't exist, but that the delusion that there is evil can be lived. In Buddhist terms: that evil has no self-being. (We shall return to this point.)

How to respond to the kind of threat that "evil" can present one with? The answer is most delicately and persuasively presented in *The Lord of the Rings*, perhaps, through the veil of the swash and buckle that is the vast struggle for Minas Tirith in *The Return of the King*. It is not sufficiently subtle, for instance, to say, as I may have seemed to say earlier, that the answer is always to go out to meet the "evil" head-on, and thus banish it. The retreat back into Minas Tirith before the onslaught of Sauron's army may seem nothing more than a mirror to the retreat back into Helm's Deep before the onslaught of Saruman's forces. The Uruk-Hai were defeated when Aragorn and Theoden rode out into them; but Gandalf's stepwise partial retreat in Minas Tirith is by contrast a *tactical, courageous* retreat. Contrast it with the irrational retreat of the Steward, Denethor, back deeper into the fortress and into himself.

Or think of the rash going out to meet the enemy of Faramir, under Denethor's twisted command.

The recommendation of *The Lord of the Rings* is: not to retreat *in spirit* (or at most, to do so under controlled conditions, as the goose at the head of a flock slips back to the slipstream of the v-formation, confident *that others will take up the slack*); but sometimes to retreat in the flesh, as a matter of good sense. The recommendation is of the Aristotelian mean: *not*, as Gandalf says to Faramir, of throwing your life away. When there is no hope, then you can create hope by going out, impossibly, as a leap of faith; when there is real hope, then going out for the sake of it is rashness, of no help to anyone, and not true courage.[65]

One who can find the mean, and can know when to go out and when not to, neither seeks to dominate the world nor is dominated by it. (And there is the golden mean, again.) Contrast that with Denethor's attitude at the time of the assault on Minas Tirith: He wonders to himself why Gandalf et al. are bothering to expend the energy they are, when they could choose *simply* to retreat, willy-nilly. He is lost within himself, almost as if he has been infected by the power of the Ring at a distance (and he surely *has*—it has *taken* effect, as soon as one is brooding on it), while Gandalf and Pippin play their part, together.

Meanwhile, the Ring (and all it represents) *naturally* brings (or even *is*) paranoia, a cruel loneliness, where the existence or possibility of others feels a torment rather than a balm. We see this at key moments in Boromir; we see it perhaps too in his depressed and withdrawn father, Denethor; we see it very intriguingly in the suspicions and fears and withdrawal into Orthanc of Saruman himself;[66] we see it wherever power has sway and wherever retreat appeals. Frodo increasingly becomes a case study in paranoia as he nears Mordor: in his lack of sleep, in his obsession with the Ring and his jealousy over it, in his distrust even of Sam. "Meanwhile" Gollum, the weak threads of his trusting connection to Frodo fatally compromised by his harsh treatment at the hands of Faramir and Sam, has by now apparently given up on human beings.[67] He has never easily been able to think of human connection as anything other than the connection between a master and a slave,[68] and now his only real connection is "with himself." His "good" "Smeagol" personality gleefully looks forward to taking "The Precious, the Ring, *for me*!" His "bad," self-protective persona responds fairly forcefully, "For *us*." The "Smeagol" persona replies, somewhat nervously, "Yes, that's what I meant. . . ." Gollum absolutely thinks of himself as two, while knowing he is one. This is the paradoxical nature of the schizoid or schizophreniform much more than are plain hallucinations, let alone than plain dissociations (e.g. the dissociations in identity found in "multiple personality disorder"). The paradoxicality of such thinking is offered to the viewer constantly through Gollum's peculiar language,[69] through the way especially in which he personifies the Ring and cannot really separate it from himself. The Jackson films even give us, in superb inspiration, a number of point-of-view shots where the point of view is *that of the Ring*. Most startlingly and eerily, when Gandalf hears the call of the Ring, in Bilbo's house, in the first film. This takes us into the mindset of Bilbo, Frodo or Gollum far more effectively than any point-of-view shots through their eyes could ever do! And, once again, this is a clue towards how the Ring is more a metaphor for the fantasy of absolute safety through absolute power than it is any real power. What *Lord of the Rings* explores is how the fantasy of finding safety through pursuing power alienates a part of one from oneself, even to the point of inadvertently creating hyper-observation (of oneself etc.) by a part of oneself. If I want to be able to watch without being seen such that I will be invulnerable, this inadvertently creates Sauron (or at least, say, a super-efficient and ever-present secret service) to watch over me. The same delusional dynamic is clearly visible in some of the most famous cases of paranoid schizophrenia, such as Schreber's and perhaps John Nash's. (It is arguably present, too, in mainstream supernaturalistic or metaphysical monotheism, of which *Lord of the Rings* can therefore be read as a devastating satire.)[70]

As intimated earlier, then, *The Lord of the Rings* concerns itself deeply with why it is so *hard*—and so slow a process—to recover reliably from deep depression or chronic obsession, and still more so from psychosis.

Now consider the climactic scenes inside Mount Doom, as presented by Jackson. In Sam's movement, in his words to Frodo, from "Let [The Ring] go!" to, a little later, "*Don't* let go! *Reach*!,"[71] Sam is presenting to Frodo perhaps the only possible *cure* to the pathologies of humanity and of reason that are psychosis. In other words: *non-attachment*—but *not* at the cost letting go of one's humanity, or of letting go of hope. Not at the price of nihilism, nor of a loss of faith in that that one loves. Non-attachment to one's cravings; but faith in, and in that sense attachment to, what actually matters: Goodness and Love; others; the Earth on and in which we exist; existence, moment after moment.

Witness here too how hard it is to live the cure, especially perhaps if one knows that one failed to live out the non-attachment part of it. (Frodo did *not* let go of the Ring. The quest succeeded only "by accident.") Witness, that is, the deeply moving close of *The Return of the King*: Frodo is convinced that he is too wounded to go on with ordinary life and to enjoy its pleasures as the other hobbits do. I am of course not denying for a moment that Frodo has gained some—indeed, *deep*—insights through his trip, his quest; in fact, on the contrary. I am not denying that it might just be easier for him to live the strange calm afterlife of the Elves rather than the "mundane" ordinary life of hobbits, as a result. But there is tragedy in his felt *inability* to go on living in the old place that his ultimately successful adventure has saved.[72] He cannot stay in his idyll,[73] "the Shire." The meaning of Bilbo and Frodo leaving on the boat to be with the Elves is that *they are not recovered fully from their psychical journeys*, or, at least, that they think they are not, which comes to the same thing. Sauron may be destroyed, but they can't forget their frailness. Frodo *doesn't feel safe feeling safe*. This is a profound understanding of profound psychological hurt and damage. (As Gollum once put it, "Once it takes *hold* of *us* [of the multiplicity that Gollum has become, and of those who are like Gollum in being viscerally subject to the Ring's power?], it never lets go.")

When everything says one is safe, one feels profoundly unsafe *as a result*, because one fears what might come in if one lets one's guard down too far as a result of trusting in one's safety. And after all, wasn't it thought that Sauron had been destroyed once before? Might he not rise again, start taking form again? Hadn't one better be ready for that eventuality, just in case? Every moment? (Thus deep sleep would [seemingly] be unwise, etc.)

If one feels any of that lust that was the Ring, if one knows that one profoundly wanted the Ring for oneself, if one feels any sense of loss that one no longer has it (and it is important to note, as Bilbo travels with Frodo to the Grey Havens, that Bilbo still feels this, too), then the Ring is *not* all gone. And still less is Sauron.[74]

This brings us back to considering, directly now, the contrasting ways one can seek to resolve the dilemma of the Ring.

- One can try to *become* the Lord of the Rings.[75] That is, one can seize the Ring to oneself in megalomania, as Sauron hopes to, and of course as Frodo does in Mount Doom (much as Isildur was in the end probably inclined to do, when he stood in the same spot as Frodo, long before). The consequences of this effort at omnipotence are obvious: one's imminent total (paranoia and) corruption is dramatized/vivified by the coming of Sauron's minions rapidly to one, in such an eventuality.
- One can simply try to hide through it, through putting on the One Ring, through withdrawal from the world, and thus seeking "safety." We have seen how futile a strategy that is, how futile and self-defeating such a stratagem of "retreat" is, through examining what Frodo's world is actually like when he puts on the Ring (on Weathertop, and similarly in the "Prancing Pony," or on Amon Hen).
- One can give the Ring away to the malign force one encounters when its power and dread (the Ring's) is upon one. What would be the consequence? A temporary relief, and then a terror far worse even than that found in the withdrawal and paranoia etc. that visited Frodo when he put on the Ring. For one would be giving total power to the alienated part of oneself that one dreaded the most. This would magnify its power vastly, and would thus mean a seemingly terminal and appalling psychosis, a world that seemed to be nothing but the rule of darkness visible and triumphant.[76]
- Finally, one can decide to go on a quest in which one confronts the demons which are each of these three possibilities, as Frodo does, and

> overcome them, as he does. (Well, at least he overcomes the second and third, with a lot of help from his friends, as well as from himself. As mentioned earlier, it is only the final chance intervention of one who is more obsessed with the Ring even than he is—Gollum—which enables the quest *accidentally* to be successfully "concluded," in the marvellous, utterly unexpected [by Sauron] gesture of giving up the Ring voluntarily to dissolution.[77] Though later I will briefly discuss how this "accident" is hardly so, when recast in terms of the "moral" of Tolkien's tale.) One can attempt, that is, to find out who one is, by seeing if one has the mettle to confront one's demons and to dissolve the haunting seductive power of the Ring completely away.

My tentative philosophical suggestion is that we see the structure of human being laid open to us in *Lord of the Rings*: to be is to be open to the vicissitudes of anxiety. But that openness is, by a (quintessentially hard) adjustment of aspect, also an openness to tranquility, to life as lived heaven (or at least, as plain life), not lived hell.

What exactly makes the difference?

In order to realize the last of the four possibilities sketched above, finally, one has always *to have faith*. Not the pathetically childish supernaturalistic version of faith—namely, belief that a certain kind of super-person exists—that is argued for (or against) in conventional theology and in standard philosophy of religion. But rather, the kind of faith explored in a religious/spiritual context from Pascal[78] through Kierkegaard to William James, and (I believe) successfully de-divinized by 20th-century philosophers such as Sartre and Wittgenstein. In Wittgenstein, in his *On Certainty* for instance, and in Stanley Cavell's work on "acknowledgement," such faith is the unavoidable faith one has in other people, in one's world, and so on. I say "unavoidable"; that of course does not mean that people have not tried to avoid having it and/or inadvertently lost touch with it—the consequences of such attempted or inadvertent avoidance in philosophy are Descartes and the various threats of scepticism, and in psychological reality are the varieties of serious "mental disorder," much as, I would suggest, explored by *Lord of the Rings* (and by other philosophical-psychological movies such as, as represented in the present work, *Last Year in Marienbad* and *Melancholia*, especially).[79]

When this faith becomes self-conscious, it is a faith *in life* which is I think life's meaning. This faith (and its failing) is all over *Lord of the Rings*; there are many examples one would want to work through, in a fuller presentation. It is explored by Jackson powerfully in his presentation of Frodo and Gollum, as discussed above. Tolkien/Jackson also famously place it in the friendships depicted in the story; notably, between Sam and Frodo. Take the end of *The Fellowship of the Ring*: as Frodo rows across the river and away from Sam, he shouts to Sam, "I'm going to Mordor alone!" Sam replies, "Of course you are. And I'm going with you."

The humour of this veils a profound truth to it. As Galadriel had said to Frodo, "To be a Ring-bearer is to *be* alone." Only Frodo can bear (t)his burden. Everyone has to do it for themselves. Each one of us *is* Frodo,[80] as we face the temptation to self-aggrandize, or to hide, or to give some power that we have away to someone else to do with it as they will. (We identify above all with Frodo, as well we might.) *And yet,* Frodo would be nothing without his friend(s).

One might venture that in effect the whole trilogy is in a certain sense shot from Frodo's point of view, even while in literal terms it quite *obviously* isn't. (It is I think important that we discover at the end that Frodo is the author of a book called "The Lord of the Rings.") We are launched on a psychological voyage with/in him. As Frodo says, "You can't help me, Sam. Not this time." Heidegger's insight that each of us has to face death on our own, and that it is in that sense meaningless to speak of someone dying for us, is close at hand here. *And yet*, once more, of course, Sam does help Frodo; in the end, Frodo couldn't possibly have done it without him, neither practically *nor emotionally*.

Aragorn and Frodo, who together have the chance of repeating Isildur's glory (of defeating Sauron), and both of whom—as Aragorn is all too aware—run the risk of being defeated by "Isildur's Bane" (Sauron's weapon, the [temptations of the] Ring of power,[81] the temptations of would-be safety *and* the consuming anxiety of self-doubt, partly *consequent* upon Isildur's example and fate, about their adequacy to their task), demonstrate what is at stake here and how one can find one's feet nevertheless, in the marvellous scene on the great stairway on the way to the bridge of Khazad-Dûm, in Moria. After all but two of the Fellowship have managed to cross the chasm that opens up on the stairway as the earth quakes with the demonic and deadly Balrog's approach, Aragorn and Frodo stand alone on a great and cracked pinnacle of rock that sways slightly this way and that. It is simply too far to jump to safety, and they seem doomed. But Aragorn sees what to do, and so Frodo, with his help, becomes brave enough to do it too. "Lean forward," Aragorn says. They lean forward, and the terrifying chasm becomes smaller—the huge pinnacle starts to sway in the direction *in which they are leaning*. They have altered its delicately poised centre of gravity. Their calm faith in action is such that finally the leap to ordinary safety is quite short. The step of faith is in the end easily made, the chasm now being much less, and thus surmountable; as their weight leaves the pinnacle, it sways back slightly, just enough such that it doesn't crush the company as it veers and crashes down into the abyss below. This is a lovely literalization of what Pascal, Kierkegaard and William James wrote about. The antidote to anxiety is faith. If you have the will to believe (and you do have the *right* to believe), if you wager *yourself*, then your faith becomes part of the process, and itself influences the result.[82] So the faith I am talking about is not of course in the existence of a supernatural person, but IN yourself, and others, and right, and being—i.e. in God, in the TRUE sense of such faith / of God.

This taking of action when there is no guarantee of success but there is a guarantee of failure if you don't believe is the essence of political struggle, of struggle in defence of "the environment," and so on. To work most effectively, it requires faith in us: faith in Being, in this flesh being *good*. (This is of course the central faith of Philip Pullman's *His Dark Materials*, a trilogy that in delicate detail exquisitely reclaims the body and its consciousness.) You can have this faith without having any empirically based hope.

Morale is all. The will or right to believe, to have faith, secularly as much as spiritually.

In short: If you *trust* that you *will* not (must not) fail, you are far less likely to fail. This is the opposite of the mental processes set in train by and symbolized by "the Ring," as discussed above. Aragorn and Frodo, the two who above all need not to become lost in anxiety, demonstrate at the powerful moment just indexed in Jackson's film how faith in action can self-fulfillingly enable one to succeed in a seemingly hopeless circumstance. In a circumstance where faith must speak for itself—a "hopeless" circumstance, or better, as we might put it, wherein there would *be* no hope, but for hope.[83]

The would-be alternative to such faith is to dose up with the drug of the Ring. The Ring is the fantasy of power. It is a drug which promises such power that one will not have to accept the frailties and disappointments of ordinary faith in others any more.

Gollum is, it would appear, more or less "hopelessly" addicted; the Ringwraiths certainly are. The *Lord of the Rings* films show others in less severe but still deeply problematic states of addiction and/or drug withdrawal. For example, Saruman when he speaks to Sauron looks like he needs his fix. Frodo of course when he starts to lose his grip, closer to Mordor, is Ring-dependent/addicted. (Sam sees this clearly, the way Frodo is changing for the worse.) Even Boromir, when he tries to steal the ring from Frodo, too.

Drugs that are addictive: What is their mechanism? They can deliver apparent safety, feelings of great capacity and power; they enable one to be (to act) normal, or "better than ever"; they offer great pleasure; they blot out. One craves them for one or more of these reasons. But one doesn't want to be addicted. One doesn't want to be a druggie. In fact, if one is hooked, one craves *not* to be addicted anymore—and, moreover, and this is the crucial point, one fantasizes, when one has recently inhaled or injected or whatever, that one is now just fine. That is to say: *the drug temporarily takes away one's sense that one is what one is (at least, so long as one's practices say that one is); namely, an addict*. One only feels like an addict when one has *not* got the drug in one. When one craves the drug, ironically, one is paradoxically craving that one should no longer be (feel) addicted.

The Ring is the perfect drug. You can have it forever—it is physical and lasting, not something temporary that will pass through your system and leave you wanting more. And it doesn't deform you—you can take it off. It's not *you*, it's not who/what you are.

Only, of course, one doesn't think of the possibility of losing it (or indeed of having it taken away from one). And one doesn't think of it changing you, or of it "becoming" (part of) you.

One doesn't think at first, that is, of its producing paranoia in you; nor of its corrupting you. One doesn't anticipate the way that one's craving for it (and one's fear that it will be taken from one) will in a vicious loop produce spiralling effects of coming to feel less and *less* safe (not more and more!), the longer one has it. One craves the "drug" or drugs so as NOT to have to face the shadow side of oneself, so as not to have to dwell in pain—the outcome is that one is left facing that fear and shadow and pain far more bluntly and terminally than one was before.

"The perfect drug" is a nightmare,[84] not a dream. *The Lord of the Rings* helps us to understand what a drug is in its essence, and it helps us understand the drug of power as a particularly pertinent instance of this; and it helps us understand the way in which (against one's intentions) the lived temptation towards unreality that is madness (and, in its milder form, derealization) is surprisingly—frighteningly—addictive.

This is partly why I stated at the opening of this chapter that I understood the primary task of *The Lord of the Rings* (the understanding of the dynamics of madness) to be psychological but also, thereby—even, in an important sense, *simultaneously*—political, spiritual and philosophical in nature. My reading of *The Lord of the Rings* is a philosophical reading that draws the psychological and the political *close together*.[85]

- Philosophical, in that, on Wittgenstein's conception of philosophy, the kind of torment that we see in *LOTR*, with its fascinating theme and variations especially with Cartesianism, *is* of a broadly philosophical nature. And also in that it is the whole nature of psychopathology that is at issue, not merely particularities of one or another form of it. Indeed, *The Lord of the Rings* perhaps even gives us some reason to think that there will tend to be more of a holism to "mental illness" than there is to the diverse set of things that are physical illnesses, for it helps us to understand the perhaps inevitable and distinctive tendency of "mental illnesses" to morph into one another. To understand, that is, how (e.g.) the dynamics of post-traumatic stress and schizoid withdrawal are very similar; to understand how both of these will almost inevitably intimately involve cycles or prolonged episodes of depression and anxiety; and so on.
- Political, in that the psychical and spiritual struggles that (I have argued) sit at the heart of *The Lord of the Rings* are importantly *mirrored* in the perhaps somewhat more obvious, surface dilemmas of power that feature in the work. Power temporarily seemingly renders one "safe"—but, paradoxically, it also renders one vulnerable, because others will want to achieve that safety. Power naturally breeds paranoia, and naturally breeds the corruption that comes of wanting to keep that power to oneself and to pre-empt any possible threat to it.

It is little surprise, then, that, as I argued earlier, what happens when one puts on the Ring is an allegory simultaneously of the corrosive effects of power and of the desperate (and counter-productive) search for a safety that one can be certain of that is the mechanism of much "mental illness." Our exploration of this mechanism, in Chapter 4, and in Chapter 2, in discussing *Gravity*, and of course so far in this chapter, can enable us to see this parallelism of politics and psychopathology which these films teach.

There is a widely shared neurosis and indeed borderline-psychosis of fear in much of the world today. *The Lord of the Rings* dissects this human trait, this unavoidable result of an unintegrated *rationality*, as much as of the irrationalities that it is normally blamed on. This helps us understand the otherwise inexplicable: how Nazis could be, as many of them were, terrified of Jews; and how Americans and Israelis can be, as many of them are, terrified of Arabs or of Muslims (cf. Chapter 1 of this book).

An ultimate challenge of *The Lord of the Rings* is to empathize not only with those who are demonized, but even with the demonizers (who of course also get demonized in turn; think Trump). I.e. to empathize with these "leaders" (and their electorates) who are responsible for the manipulative politics of fear. They—their actions, their policies—should be severely criticized and intelligently opposed; and they should be empathized with, too.

Remember the critical moment in *The Return of the King* when Gandalf et al. come to realize that Sauron is *afraid*. Far from all-powerful, then. Rather: afraid.

This is the beginning of empathy.

And *The Lord of the Rings can* foster such empathetic understanding. Not of course through what is on the surface, at its end: Tolkien's own pitifully inadequate, out-of-date resolution of the problem (namely, after the destruction of Sauron, the restoration of "the true king"). But through understanding that Sauron's dilemma was also Gollum's, and Boromir's, and Frodo's. We have failed in the imaginative task that *The Lord of the Rings* sets us, then, if we fail to feel love for Gollum, and Boromir, and even Denethor—and even Sauron, insofar as he exists. (Much as Frodo's quest almost fails—Gollum almost becomes a temporary lord of the rings—because of Sam's and Faramir's failures to entertain the possibility of *love* for Gollum, their failures to *identify* with him enough.) And if we fail to notice, for instance, that the irrational saving of Wormtongue's life reaps at last a reward: in *The Return of the King*, the men of Rohan appeal to Grima "Wormtongue" as a man, as a human, as one—as part?—of them. *This* helps him to fight free of his quasi-drugged vassal-hood to Saruman, and finally to strike a blow for good. Aragorn's faith in humanity, his not having given up on anyone, is rewarded.

Just as Frodo's faith in humanity, his not having given up on anyone, is rewarded, when Gollum not only leads him into Mordor, but does him the kindness of taking the Ring to its grave in the lava within Mount Doom, too. Though these rewards were not of course even remotely foreseeable; for then they might have prompted more or less corrupt goal-oriented action to realize them. Not giving up on others *only* because such is a reliable route to getting what one wants as an outcome is a Ring-like strategy, and not true virtue at all.

We should understand then that in the end Sauron is actually an alienated part of ourselves. He is the monster that we ourselves create, and turn over-and-against ourselves. We must understand this, again, without of course *indulging* our inner Sauron (especially if that Sauron gets externalized, into a powerful ruler). Rather: we pay attention to it. We stop it growing and taking dominion, by facing it, with at least a little love in our hearts.

Sauron is Stalin, he is Saddam Hussein, he is Bin Laden, too, insofar as each of these was a reactive product of the fears and injustices of "the West." He is also Trump, and Trump's "base"; he is in more general terms our *dreadful* leaders in the contemporary West, who some might well argue are the leaders we collectively deserve until we heal ourselves and can bear truly to look at ourselves. If we re-elect them when they terrorize us, then they cannot be wholly blamed. If we even just partake of any of the same impulses moving them, then they cannot be othered.

We are living literally in states of terror, and it is difficult to see how the interminable "War on Terror" (i.e. war *of* terror—for its weapons, from torture to "daisy cutters," are terror-weapons, and its product is a permanent state of semi-suppressed terror on the home front) will ever end. Because, as *The Lord of the Rings* teaches, feeling safe—lowering your guard—is when you are (or *feel*) most vulnerable. The overcoming of the "War on Terror," a "war" which tries to ensure that we will all be terrified indefinitely, is a task of faith and hope that requires *inter alia* the delicate task of transfiguring the *parody* of faith that is close to the heart of modern American electoral politics (not to mention the "faith-based" intelligence methodology that seems, notoriously, to have underlain assurances of Iraq's WMD capacity).

The clash of civilizations thesis is of course a disastrous self-fulfilling prophecy, an inversion of cause and effect. The clash of fundamentalisms—both sides fighting for "good" and against "evil"—is by contrast alarmingly close to the truth. The fundamentalist's credo is pathological, a seemingly total quasi-psychotic immersion in the mythic battle of Good and Evil.

The schizoid is at least trying to figure out the strange character of his "world"; whereas the fundamentalist has in effect passed through (our fantasy of) a psychotic break, and lacks doubt about the world. The "madman" who worries about being mad is at least closer to giving up the feeling of stuckness than the madman—the soldier or leader (or, great danger, both)—whose mind is never crossed by such doubts, whose delusional system seems complete. The "madman" is closer to seeing: that you are fighting nothing (or, at worst, Nothing). That we're all on the same side. This is the cure: not to fight but to lay to rest the Othering and the desperation for safety that caused one to start fighting in the first place (as with the primal scene of

Smeagol and Deagol). To take the necessary risk of allowing oneself, at least usually, to feel safe.

The way *Lord of the Rings* (*especially* the film version of the trilogy) is usually seen, then, as a titanic military struggle between Good and Evil, is, on my reading, *the very pathology it (the film) is trying to cure*. One *must not* allow all the swash and buckle of Jackson's creation to obscure from one's view what the story actually makes clear, even on its own terms: that it is essential *not* to take up and use the enemy's weapon(s). The genius of those who saw what must be done with the Ring, including even men like Aragorn and Faramir who overcame themselves in order to see it,[86] is the indirect route to peace. One must not even try single-mindedly to *gain* peace (for that is the same logic as putting on the Ring); one must simply let go the weapons of war, completely. For again, Frodo's quest was an astounding one, never dreamt of by Sauron:[87] to take the Ring—to take (the fantasy of) non-cooperative and violent "total" power as a solution to anything—back to its starting point, and lay it definitively to rest. To dissolve it away. Then Sauron ceases to exist. This should be thought of not as killing him, but as dissolving "him." As laying to rest a fantasy that we have set up over-and-against ourselves, a fantasy that is lived so long as one does not have the bravery to set it aside. *Not* to oppose it head-on—that would be in the end quite hopeless, as Gandalf most clearly realizes. The final, awful, hopelessly doom-laden great battle before the Black Gate *willingly* undertaken against the forces of Mordor is of course only a deliberate distraction, a *diversion*, to enable Frodo to achieve his quest. (It will be hard enough for him to achieve it even without Sauron's gaze upon him.) They go to battle, this time, to lose, but, through distracting the enemy's gaze, still to win. To win, you must lose.

All this might in turn be best thought of as a metaphor for meditative practices that involve giving up the fantasy of self (or also perhaps for the Alexander technique). For the answer is then *to dissolve the enemy's weapon*, by dissolving (in the psychical fire that birthed it) the fantasy of achieving safety through withdrawal from the shared common world and/or through coercive power against others in that world. *The Lord of the Rings* does *not*, superficial appearances to the contrary notwithstanding, concern victory in battle; it concerns the dissolution of the fantasy that one can truly win through violence, through secret/powerful weapons.

Much as Polanski's deep and dark *Macbeth* ends with a returning Donalbain hearing the witches' siren call, and we know that the whole cycle of ambition and violence will likely recur, so the restoration of the "true" or "good" king at the end of *Lord of the Rings* has not solved the problem. It has not even much mitigated it. Frodo's great effort, his supreme non-violent struggle to dissolve the violent power of craving for power, in which he succeeds (albeit in the end by accident; the Ring is too much for him, but fortune in the shape of Gollum's craving for it leads to its plunge to destruction in Mount Doom),[88] may have been in vain. Just as Donalbain, the younger brother, is liable to try to seize the throne in the future according to Polanski's *Macbeth*, so perhaps Faramir (Boromir's younger brother, and the heir to the stewards of Gondor, and in alliance with Eowyn of the Rohirrim) might one day try to seize it from Aragorn, if and as Aragorn threatens perhaps to become a new Sauron, a new tyrant.[89] So long as the throne exists, so long as it is the sword that undergirds it, and so long as implied violence is sovereign, then the Ring in practice *still exists.*

Scenarios such as this are *felt* by every king ("true" or otherwise), and lived out horribly by some madmen.

But, we can also experience the passing of the Ring. Such a scenario of gnawing corruption, paranoia or fear can be seen in a totally different way—that is, by omission, by glorious contrast—on Frodo's face, when he receives the inconceivable liberation, the blessing, of finally being Ring-less, after the Ring is at last dissolved away in Mount Doom. It truly doesn't matter to him then if he lives or dies—he is in paradise; in the present. All is well. "I can see the Shire," he breathes, smiling.

If there are echoes here of the positive psychology and of the non-violent ethics of Buddhism, that is little accident. What Buddhism (and Gandhi's philosophy of non-violence, and Marshall Rosenberg's Non-Violent Communication, and arguably any true spiritual path or contemplative and yet engaged approach to life) reminds us of is that we are one. We are never sundered from others. This inner realization needs however to be lived, if it is to be realized. It needs to be bent to creating a new Earth. The task that will take centuries, at least, and that will require that one day all the swords that we see wielded by the ostensible heroes of *The Lord of the Rings* be turned into ploughshares, is a task that will most naturally come to seem pressing and possible in a world which has overcome the madness of (belief in the existence of, or fear of) God and lord/king *alike. Lord of the Rings* then suggests a future path beyond that linked pair of illusions. Buddhistically, according to my interpretation of it, it does not take Sauron to be something out there. Like the Mara, Sauron is our creation. Sauron is our own failings, our own temptations. To reverse Nietzsche's figure (when his madman shouts that God is dead): We have *created* him—you and I.

Shunryu Suzuki, the great Zen master who brought Buddhism most influentially to San Francisco, used to speak of one being "burnt up" by Buddhist practice. That there should be nothing left of one. When we think of the self-sacrifice of Frodo and Sam at Mount Doom, where by rights they should have been burnt up (as Gollum was),[90] we might think of this as a literalization of the Buddhist idea of renunciation. Indeed, as David Loy and Linda Goodhew argue in a very useful piece, "The Karma of the Rings: A Myth for Modern Buddhism,"[91] they could helpfully be seen as bodhisattvas: beings who prolong their own suffering, and give up their self-ambition, to ease the suffering of others.[92]

And they do so, I argued above, essentially non-violently, or at least for the sake of non-violence. It was Bilbo's pity towards Gollum (in *The Hobbit*), and then Frodo's repeated self-same act and attitude, that made the triumph at the close of *The Return of the King* possible. Here is Buddhist ethics.

The world of *The Lord of the Rings* films is not a godly world. It is by contrast, as I have argued, a world decisively devoid of supernatural religion. This fits it for being seen in the light of Buddhism's non-theism. And its demonology is of course compatible with the same in Buddhism: the demons that Buddha encountered and that we can encounter in meditation are predictable products of delusion. They do not have an independent existence—evil lacks a self-identity in Buddhism and, I have argued, in *The Lord of the Rings*. Compare here the following remark from one of Tolkien's letters, concerning his trilogy: "I do not deal with Absolute Evil. I do not think there is such a thing, since that is Zero."[93] Here Tolkien and Buddha stand with Socrates, St. Augustine and various other distinguished personages, in resistance to most Christian orthodoxy, to Manichaeism etc. (Evil is our desire to be mini-gods: our desire to have "rings.")

But again this does not of course betoken an overly accommodating or "relativist" attitude towards the very real harms that people do to one another in our world: far from it. In the post-theological world of Buddhism and of Tolkien/Jackson, the responsibility is on us all the more firmly to love each other, and save the future, in the face of the enterprises of our collective Sarumans. As Loy and Goodhew put it:

> Hobbiton expresses Tolkien's nostalgia for the vanishing rural England in the West Midlands of his youth, but we should not dismiss such homesickness with the reassuring Buddhist maxim that "everything passes away." Our collective attempt to dominate the earth technologically is related to the disappearance of the sacred in the modern world. If we can no longer rely on God to take care of us, we strive to secure ourselves by subduing nature until it meets all our needs and satisfies all our purposes—which will never happen, of course. Because our efforts to exploit the earth's resources are damaging it so much, the fatal irony is that our attempt to secure the conditions of our existence here may destroy us.[94]

This is a well-put reminder of how an ill-judged "Promethean" quest for security can be self-defeating,[95] in the very same way as the mechanisms of withdrawal and of lessening trust in others self-defeat. Loy and Goodhew go on:

> Is there a clearer or more dangerous example of institutionalised delusion? We are one with the earth. When the biosphere becomes sick, we become sick. If the biosphere dies, we die. The technological Ring of Power is not the solution to our problems. It has become the problem itself.

The Lord of the Rings calls us to go towards our demons. To face them. And then to see them dissolve.

This call *is at the same time* a call to action to defend—and restore—the ecosphere, including crucially against the dangerous dreams of technophilia (that I discussed *The Two Towers*'s critique of above, and that was heavily critiqued in Chapter 5, and in the coda on *Avatar*, below). For the truth is even more direct than Loy and Goodhew put it here: The ecosphere becoming sick IS us becoming sick. Its state is an externalization of our inner states; and there IS no "it" and "us" in the first place; we are *nothing* without it. It is part of us (and vice versa). It's a gigantic and dangerous illusion to think that we have any existence *at all* apart from it.

This is engaged Buddhism in the 21st century; and Tolkien's vision, as visioned by Jackson et al., is an apotheosis of it.

Randel Helms remarks, usefully, that "The poetry of the mythic imagination will not, for Tolkien, replace religion as much as *make it possible*, putting imaginatively starved modern man once again into awed and reverent contact with a living universe."[96] *The Lord of the Rings* is then a spiritual work without being a supernaturalistic theistic one. It seeks—it offers—philosophical, spiritual and (dare I say it) practical remedies to philosophical, psychological and political maladies. In placing us eco-politically back in contact with a living world it might, like *Avatar*, make a spiritual/religious orientation possible again. One of a partly new kind, but based in the very old.[97]

Lord of the Rings takes the Hobbesian problematic in political philosophy (the problem of mutual mistrust and mutual contract), it shows the disastrous consequences of an attempted Hobbesian solution (a collapse of trust, paranoia, the possibility of a brutal war of all against all "salved" only by a brutal—or, one fantasizes, by a benign—sovereign) and it suggests (what I would suggest is) a strikingly un-Hobbesian solution: unconditioned love, democratic and unconditioned mutuality, *never to give up on anyone*. And implicit in this is a key lesson of what I have called this allegory of madness: that there *is* no complete madness. That the only madness is to believe, as we are endlessly tempted and encouraged to do, that there is madness that one can get definitively lost in. That the only evil is to believe that there is Evil. To believe that there are people—mad people, or Evil people—who must be wholly given up on. To believe that there are human organisms who are not human beings.

It is the midwifing of the learning of how to overcome this belief that is the task of this text. Considered as an epic with an exciting plot, *Lord of the Rings* is a spectacularly disastrous failure, its plot flaws and implausibilities (and its lack of real suspense) too drastic and numerous to enumerate. The most basic of all these plot flaws is that the giant eagles who conveniently appear at two key moments in the story could simply whisk Frodo to Mount Doom to chuck the Ring away into the lava of Mount Doom lickety-split as soon as Gandalf realizes what the Ring is and gets hold of the eagles to help. The whole story would be over before the Fellowship even needed forming. But the Fellowship is formed—and indeed Boromir is allowed to be and remain part of it even though he will surely try to take the Ring; again, this is fairly absurd, not believable. And the absurdly torturous journey on foot to Mount Doom is undertaken, because this is of course a quest in which Frodo—and Boromir,[98] and you and I—is exploring *himself*; and developing his (our) ethics. It is—of *course*—a shared but *psychical* quest.[99] It is a psychological journey we watch and viscerally experience, a journey of and about Frodo's willingness to journey into his mind and heart and to be willing to sacrifice even his sanity to show a way beyond the fantasy of safety through force, through "power."[100]

The intricate plot, *Lord of the Rings*'s story, *in no way* then explains its vast appeal, *its* power: to reach and move and haunt and compel audiences. Rather: *you* have to go through this (deliberately difficult, intricate and *epic*) trial; *you* have to come back from death and from a "fatal" withdrawal; *you* have to forge relations of trust where it seems quite impossible to do so; and so on. These tasks are yours, simply because you are human. *The Lord of the Rings* is above all a virtual psychical test that one takes oneself on, and an education.

Lord of the Rings intimates a quasi-Buddhist "middle way" between an absurd and self-defeating idea of one's dominating the world, and a defeatist way of one's being dominated by it. It shows that these are in fact the same idea. The Aristotelian "doctrine of the mean" translates here into neither rashness (Faramir throwing his life away for his father) nor quietism (Denethor retreating both physically and psychologically,[101] at a loss to understand why his subjects are bothering expending the energy trying to defend themselves), "even" when one's self / one's community / one's entire world is under dire threat. And thus it is that Aragorn speaks of "this good Earth," in his last speech. We are, I believe, badly in need of the ecological consciousness, the "green therapy," of *The Lord of the Rings*, at this moment in our planetary time. We desperately need people, now, who will not retreat from this world, even when it is under dire threat. Rather—*especially* given that it *is* under such threat.

There is an alternative kind of safety available to that which we fantasize as our need when we dream of seizing power either in the public realm or in the "private" realm of our own minds.

It is available when we take "refuge" in the Earth, and in each other. And when others, and even the Earth, simultaneously *take refuge in us*. That is the paradoxical, true teaching of this work, I believe.

All of those who we encounter in any relevant fashion in Jackson's films are tempted by the Ring. Gandalf is tempted by it in Bilbo's house; Galadriel when Frodo offers it to her; Aragorn just for a moment as he hears the Ring calling his name, its barely audible siren call to him, when Frodo asks him whether he (Aragorn) can really trust *himself*, can really save Frodo from himself (Aragorn); Sam just for a moment soon after rescuing Frodo in the Tower of Cirith Ungol (the world and ordinary voices fade out, as Sam perhaps feels "Ring-world's" presence for the first time; before Frodo tells him that he [Sam] is not up to carrying the Ring, and re-assumes the burden); and so on.

A possible morally corrupted near-omnipotent Galadriel, say, of whom we get a glimpse through Frodo's eyes at their magnificent private meeting, might remain physically beautiful, but would be a secretly terrified creature too, an alienated paranoid being without any confreres, dishing out whatever it took to retain her position, insisting on the "love" of all for her, requiring that she be "worshipped" and "adored."

The fantasy of safety through absolute power is the shared *telos* of projects of violent elite rule and of projects of withdrawal to a place of absolute mental suzerainty. The pathological instability of the former is demonstrated, I claim, by the pathological instability of the latter, as clearly witnessable for example in the writings of Daniel Paul Schreber. Writings such as his provide one with raw materials which can be psycho-philosophically dissected in such a way that their inherent tendency to oscillate between a sense of near-omnipotence and a sense of powerlessness to the point almost of non-existence, and to be haunted by others including most probably a near-omnipotent Other (however improbable the actual existence of such rationally seems to one—for one is in a Pascalian situation here; dare one wager against the existence of a malign demon, who could and would gleefully torture one for eternity?), becomes perspicuous. I have done some such "dissection," through *LOTR*, and referred to more, here.

***

I must deal finally with one especially pressing objection. It will be claimed against my reading that I can hardly hope to find the emancipatory and loving message I see in and beyond *Lord of the Rings*, given that this is allegedly a racist text. How can it yield a (post-)Christian, quasi-Buddhist call never to give up on humanity, and indeed to overcome the myths of theism and even of monarchy and the like, if it is half-founded on a xenophobic hatred of the "Eastern races," of the dark-skinned etc.?[102]

I have already hinted broadly at my answer. Yes, Tolkien's text is unfortunately an allegory troublingly tinged, at times, with racism/speciesism, and yes, Jackson's films fail to seize the opportunity to overcome this unfortunate legacy (some tribes of the Elves, for instance, could have been black, which might well have done the trick; it is perhaps regrettable that, while Jackson seized the opportunity to inject a healthy extra dose of feminism into Tolkien's story, he failed to do the same with anti-racism); but the counter-argument, the defence which my interpretation makes available, uniquely, against the charge of racism is this: Sauron and all his minions—the Black Riders, the Orcs[103] and so on—do not really exist. They exist only in Frodo's mind (or in your mind and mine). This is what the insides of our minds sometimes look like (especially perhaps in the West): mostly inadvertently, we populate them with what are to us horrifying monsters. With our own alienated selves, and with fantasized unreachable others who must be given up on (which is what we fear that we ourselves are, or could become).

The Ring-wraiths are neither alive nor dead; they are impossible phantoms. They are our nightmare, nothing more (and, admittedly, nothing less). Even the great armies of Saruman and Sauron may be seen as really just devices of dread.[104] Either they are not totally lost; or, those that are, such as wraiths, do not exist except as 'summoned' by us. When one retreats into one's mind from a hostile world, one is still—in fact, far *more*, as Frodo demonstrates to us—vulnerable to the awful vicissitudes of the non-existent: of ghosts, demons, wraiths. What you see powerfully reflected, when you come to understand your mind better—and I am suggesting, following (or, if you prefer, anticipating!) Laing and other astute psychological commentators on this, that it is this self-knowledge that *The Lord of the Rings* gradually generates—is (y)our own alienated craving and destructiveness.

Thus one need find little or no racism or speciesism in *Lord of the Rings*; rather, see the encounter with it as a "meditative" experience where you allow these wraiths and phantoms and "dark" lords and goblins to come to you, and to pass away. (This is a positive—therapeutic—alternative then to being *caught up* in these objects of fear; that way, madness lies?) One doesn't give up on any human/being, one doesn't lose faith; and all that is given up on then are phantoms, fantasies. Not only is Tolkien's malicious demon far subtler and more psychologically realistic and more terrifying than Descartes's by virtue of being incomplete without *you*, but he (Sauron et al.) *does not exist* except insofar as you will him to, allow him to, and so forth. In Heideggerian terms: "he" is ontological—an aspect (or not) of your lived world, entirely interdependent with you—and not ontic. *The Lord of the Rings* on my reading, then, is a very non-standard, Milarepian[105] demonology: "virtual," imagined terror is being fomented (by oneself, or by others for one) in this story, *just as* real terror is being fomented across the Western world, today. Principally, by right-Populist governments such as Trump's, and by their more or less willing or sleepwalking accomplices in the media, the electorate etc. This terror is "splendidly" self-fulfilling: following almost precisely the dynamics we have seen in the films discussed in this chapter, this terrorization of countries by their own governments (and of people by themselves) is (ironically?) "self-verifying," in that it continually stokes the very hatred of the West, the very malevolence, that was initially projected onto the East (and South). It produces the very malevolence that it aimed to put the lid on, to make safe, to achieve safety from. And then it claims that that malevolence was there all along; and thus it justifies itself.

It is as in Adam Curtis's powerful BBC TV series, "The Power of Nightmares": one constructs an enemy to deceive one's population and thus gain a greater power over them, or because one is paranoid, or both. Your construction becomes a self-fulfilling reality; the measures you take, of vilification and repression, produce a reality not dissimilar to that claimed in your propaganda; you can then claim vindication for your propaganda as true!

If and as your enemy gets weaker, you have to propagandize *all the harder*; you have to look harder for threats, and "build them up" more, the stronger you are.[106] This "looking harder" has a precise analogue in schizoid conditions (and in derealization more generally). Much schizy hallucination is actually not the seeing there of things that are not there that is paradigmatically imagined in our popular images of madness; rather, it is the result of straining hard to see *whether* there is something there that is abnormal. If you stare hard enough, your perceptions *will* come to take on a strange quality. You will then take yourself to have been vindicated in your watchfulness. (Perhaps you will then forbear to go to sleep, so as to be able instead to watch all the harder and more constantly. "Surprisingly" enough, you will then see even more that confirms your worst fears. The price of eternal vigilance is becoming terrified that the slightest glimmer of a threat will take away your liberty. This is the actual state of the U.S.A. [and the U.K.] today: Liberty is being pre-emptively sacrificed on the altar of eternal vigilance.)

All this needs to be understood, so that we can work with the state of mind of our fellow citizens and our politicians. So that we can understand, and not condemn; or, if we condemn, then we must condemn policies, not people.

All this needs to be understood so that we can hear and understand the part of *ourselves* that would give up our freedoms in the name of security or convenience. The remaining task that *none* of the protagonists in *Lord of the Rings* really engages with is (then) surely this: to learn some love for any actually existing outer Saurons *and* for one's inner Sauron, as a grandmother loves the children she gently watches over, and does not regard any of them as entirely mad, bad or dangerous. And, further, to learn to listen to and not respond with violence to the inner Saurons of others. It might not be enough, to dissolve the Ring to nothing. It's only a piece of metal, a symbol.[107] If one does not embrace and thereby perhaps eventually dissolve to nothing the part of oneself that the Ring appeals to, then one will not surely have completed the quest. One needs to give up of one's own accord the tendency to magical thinking that is represented by this bit of metal, the tendency to think that one can achieve something marvellous through paranoia and the like. One needs to smile at it, and let it dissolve into nothingness in the beautiful fires of oneself (and molten lava is not just sublime, but sometimes *so* beautiful).

The quest, I've argued, is for a kind of wholeness—a wholeness in and of oneself, and the wholeness (or oneness) in and with society and our ecology that is its necessary flipside—that is not dependent on the pathological nature of one who is not whole without the Ring.[108] And so long as the Ring exists—and its existence, I am now suggesting, could hardly depend on the mere existence of one particular bit of metal—any being that comes near it is not quite whole without it.

My reading can thus understand what is otherwise something of a mystery: why the title, *The Lord of the Rings*, is not only powerful but apposite. One might think that the book would have been more accurately named simply *The Ring*, or *The One Ring*. These titles would however miss the point: that, in the end, the Ring is literally nothing more than a symbol, a cypher. What matters is the psycho-socio-political dynamic that it stands for. What matters is the *dread* of the Lord of the Rings and the almost coterminous desire to *be* the Lord of the Rings; and the understanding of what is actually involved in dissolving that dread and giving up that desire. That is why the title is so right.

We can now state my non-standard reading of *Lord of the Rings* even more strongly than we stated it at the outset, when we considered the corrupting effects of power: it is not precisely power that corrupts, but rather the desire for its achievement or continuance . . . or, more precisely, the unstable nature of its *always* fantasized achievement. This is how psychopathology and politics are *never* more than a hair's breadth from each other: the truth of the wish for conventional political power[109]—the power in effect of kingship, of lordship, the kind of power one can allegedly "seize the reins" of,[110] and ride—is expressed vividly and clearly in the withdrawal and drive towards quasi-solipsism of most "mental illness." Sending your minions out to conquer the world *is* a kind of paranoid withdrawal. The drive towards total power, towards a fantasized complete overcoming—a full spectrum dominance—of the "slaves," the insignificant creatures who might threaten one, is an all-too-human futile denial of contingency, vulnerability, intersubjectivity. It is a denial of the need to trust, and of the willingness to expose oneself to what Tolkien calls "the world of men." It is soulless. One is strongly inclined to call it insane.

And one should understand then that Sauron and the Nine wanted the Ring for the very same reason that Frodo and Gollum and Denethor and Boromir and even Faramir and Gandalf and Galadriel and Aragorn and Sam wanted it, for themselves—it is only human, to want, impossibly, such safety as it would seem to deliver.[111] If my reading is right, *Lord of the Rings* might at last then truly be absolved of the criticism of subtly or unsubtly fomenting racist hatred, through the defence that the objects of the hatred are merely psychical.

And of course, it is in the end of relatively little significance whether my interpretation *of the text* is right. What matters is less what I say about the art that Tolkien or Jackson made, of the "world" that they created perhaps without any thought of its interpretation, of it as "allegory"; what perhaps matters rather more is the substance of my analysis of politics and psychopathology, my philosophical and ethical claims. This chapter on *The Lord of the Rings* has inevitably been selective; I certainly do not claim that *The Lord of the Rings* is once and for all only this allegory. Rather, I mean to have illuminated an aspect of this great myth. This chapter about this new myth is itself perhaps (like) a myth.[112] If it is, then I do not believe that it is any the worse for that.

Tolkien/Jackson aim to re-enchant the world: a timely project, for this is a quintessential neo-myth for our times (as I will explain further in drawing on a broadly complementary neo-myth, *Avatar*, below).[113] I have emphasized that, in a world where the myth of God is half-dead, there remains something unavoidably, mythically powerful: the *fear* of God. Or rather, the not-unreasonably morphing constellation of fears and desires that I describe as the Ring and its Lordship: the fear of (the wanting to be)[114] the Lord of the Rings, the fear of his punishment of one, the sense of being watched.

What finally matters, perhaps, is—to complete the cycle of overcoming hatred and violence—that one must understand and embrace and not hate the "part" of oneself and of others that hates, and lusts for power, and that one hates. *We are one*. In the end, it isn't the Ring that binds together; our preciousness and quest for unity even in (or especially in!) utmost adversity *is* our binding. (This isn't merely a dream. To assume that it is is already faithlessness.)

If one—if we—can thus become who one is—who we are—*whole*, then there is a chance for *realizing* the wholeness *of all beings* that perhaps lies at the base of the ecological vision of Tolkien, as of the Buddha. You, me, Donald Trump and Gollum: We are one.

You and me, all of us, precious.[115]

***

Avatar begins with a dream. A dream of flying[116] over a beautiful place with a working ecology. We might call it: a dream of the Earth. A dream of what the Earth has been, and could be again, if we restore her wholeness, her ecosystems.[117]

The opening sequence is quickly followed by a long sleep and an absence of dreaming. In a voiceover, the protagonist (an ex-Marine everyman,[118] *our* avatar *in* the film) says that if you are cryogenically suspended, "You don't dream at all."

With this double movement, the film seems to suggest that in a certain key sense we (the audience/viewers) don't dream *enough*. That there is something sad about a state of being where one is alive but not dreaming. Where one has all but given up hope: for one no longer has a dream. The voiceover describes a state of being created by the hyper-technological world that the Earthlings in the film come from (an extension into a possible future of our *actual* world), a world obsessed with extending life (e.g. cryogenesis), with escape from the Earth, and with sucking the planet (any planet) dry of minerals for profit.

In contrast with these attitudes/behaviours, the film seems to be inviting us, the audience, to dream. *Avatar* suggests that to dream is to *live*; that to live without dreaming is not really to live at all. That if we live in such a way we might as well already be "in cryo," frozen. In other words: the "you" in the above quotation is *you*, me, us.

From the start *Avatar* suggests simultaneously that we need to wake up and open our eyes, to really see things, as they are; in order, perhaps, to become enlightened. In this regard, *Avatar* follows of course a venerable "philosophical" tradition of "enlightenment" films. Another particularly striking case (besides those analysed in earlier chapters of this book) is *Fight Club*, and its climactic line, just prior to "Jack's" overcoming of his parasitic "Tyler Durden" self, "My eyes are open." Compare also the meaning in context of the replicants' calls to Deckard in *Blade Runner*: "Wake up!" *Avatar* pulses through a lengthy sequence of fallings-asleep and wakings-up. That is in effect how the humans switch in and out of their avatar personas. It is a time-honoured and well-exploited metaphor for enlightenment, for truly being alive.

Take for instance Grace's remark, to Parker Selfridge (the corporate boss figure), one of many similar examples in the film: "You need to wake up, Parker. The wealth of this world isn't in the ground, it's all around us. The Na'vi know that, and they're fighting to defend it." This is what the film tries to teach, about *our* world. Can it arouse in us sufficient awareness, to really *see*, hear and (most importantly) act?

In the opening scene of *Avatar*, the protagonist's eyes are closed (he is in cryo), but the film ends with a shot of a pair of eyes—the same eyes, and yet not the same—opening. We'll come back to that.

The indigenous beings in the film, known as the Na'vi, call the avatars in which earthlings walk on their home-world, Pandora, "dreamwalkers."[119] This seems to be a desperately apposite term. So long as the avatars remain aloof from the lived social and ecological reality of the world in which they find themselves, so long as humans fantasize about being separate from their (our) world, they are—we are—as if dreaming, because asleep. We might as well just be dreaming because an illusion exists that comes between our *actual* reality and our *lived*, or *assumed* reality, and it blocks the former out. In this way the human characters depicted in the film are *merely* dreaming; as opposed perhaps to deeply or great-heartedly dreaming, dreaming like Martin Luther King did. We might say they need to wake up and discover that the dream, the beauty, the hope is *real* (and fragile). As the ex-Marine, Jake, puts it: "Everything is backwards now. Like out there [on the living surface of Pandora] is the *true* world, and in here [the artificial atmosphere of the Earthlings' colonial outpost] is the dream."

So: we don't dream enough. Yet, as we saw the opening of *Avatar* already telling us, we do fantasize. *That* is what we have to wake up from (doing). We seek to escape reality. (And *this* was the key, we saw above, to *LOTR*. *Avatar* counterposes dream with mere fantasy. While at the centre of *LOTR* is nightmare, *resulting* from attempted escape from reality.) We are ungrounded. *Avatar* suggests we live a dream *in the wrong way*. We need to awaken (from this errant form of dream). And if we are to awaken, presumably we will have to learn to dream properly again. We will have to dream of restoring our planet and radically change our values. We will have to learn from indigenous peoples, in much the same way that Jake does.

Pandora is a (dreamed-up) world of miraculous beauty and wonder. *Avatar* of course utilizes 3-D special effects, encouraging viewers to enter into this wonder-full world as deeply as possible. Total immersion, visual, auditory and kinesthetic (seeing, hearing, *feeling*), is the aim. We, the audience in the 3-D cinema, mirror Jake, who needs to truly and fully experience Pandora, to drop into his avatar, and enter into the world of the Na'vi completely, just as the Na'vi connect deeply with each other, with their animals and their surroundings.

3-D technology makes it possible for viewers to feel as if they really are there, which is crucial to the success of the film's meaning and "message." One has, I shall claim, to feel Pandora as if it is real.[120] Because it kind of *is*. It is *our* world, seen through 3-D glasses darkly. Or, better: it is *our* world, through a glass brightly. The film plays with one's sense of reality. *You have to come to feel it as real. Not*, as I discuss below, as if it were a video game, but by opening to the reality of *our* world. By opening us to the perfection of the world and the perfection of humankind in full awareness, which truly enables us to see the Earth, our home and one another, *face-to-face*.

There were numerous reports of people becoming depressed after seeing *Avatar*[121] because reality isn't as beautiful as the world they had been inhabiting. Though much depends upon what world one IS inhabiting. A colleague of mine showed *Avatar* in situ to a tribe of Amazonian "Indians." None of them were depressed after the film. They commented along the lines of: we live this. What you have just shown us is our beautiful, threatened place.

I think that the experience of leaving the "movie theatre" after seeing *Avatar* in a jungle must be profoundly different from seeing it in a mall or on a high street. So, let us ask: What causes "*Avatar* depression"? Is it the contrast between reality per se and Pandora (which would suggest mere depressive escapism); or is it the contrast between conditions on (too much of) the Earth *as we are living it* and Pandora? The latter would be my suggestion. In other words, whether they are aware of the fact or not, I submit that in all likelihood such viewers might be depressed about the ways in which we have despoiled our Earth, ruined a Paradise (and of course this despoliation shows little sign of abating). The world we inhabit is often ugly because humans have made it so. In other words: viewers may feel depressed by the very thing *Avatar* wants to render focal and aims to change. Depression is one possible—and natural—reaction to what we have collectively done to the world (and by extension to ourselves), but a healthier reaction would be to turn depression into anger and ultimately into the will to forge necessary change. This can be achieved by the transformation of depression into an ecological consciousness.[122]

Compare for instance this account:

> Anna, a young woman who cried for an hour after watching [*Avatar*], told me about her experience: "The feeling I had was one of mourning: mourning our loss, as a species, of our connection to the basic sustenance of life. . . . *Avatar* has contributed to a growing ecological

consideration within me; I am finding it increasingly difficult to assume the position of a lack of personal responsibility by the 'burying-my-head-in-the-sand' method."[123]

Is this kind of thing just sentimentalism? Not at all. It seems to me a profoundly sane response to the escalating ecological challenges we face. To the charge that *Avatar* sentimentalizes its world,[124] we can in any case reply in the negative, for Pandora is a world more naturally hostile than ours. (There are no fluffy bunnies.) Setting aside for a moment the poignant fact that humans cannot breathe the atmosphere, that the air itself is death, the Pandoran jungle is a morass of menace. Our hero is contemptuously—and repeatedly—described by his love (before she loves him) as being "like a baby." He has no clue how to survive in *this* natural, hostile world, an irony given that Marines are taught fundamental survival techniques. Similarly, so many of us now find ourselves at sea in nature;[125] and this will become an increasingly serious problem if we are forced to rely on the Earth . . . as we will be.[126] We tend to deny and fear nature while simultaneously clinging to hopes of techno-fixes as solutional to our environmental, economic and even our spiritual problems.[127]

Avatar, like *Lord of the Rings* (think especially of Isengard), suggests various limits to techno-fixes. In particular *Avatar* suggests that the techno-fix mentality involves at its core an evasion and a loss. An evasion of true human nature, and a loss of the sense of beauty and connectedness this can yield.

Of course, being "like a baby" can have its upsides—if one needs to see things genuinely afresh, to be reborn, to be looked after and taught by a wise, protective parent. We, the audience, observe Jake "growing up" through various stages in *Avatar*. Repeatedly our hero is forced to shed the layers of his well-constructed masculine ego. In a sense Jake needs to die to his former life in order to secure a healthier future—for himself and others. (This creates an interesting twist on what it means to be a "serviceman." Marines are members of an elite force. Their purpose is to defend the vulnerable and to willingly sacrifice themselves for the greater good. In our world, sacrifice is almost always connected to physical violence. But in *Avatar*, the most important sacrifice Jake must make is to release his ego in order to be reborn. This is the higher service.)

Here's how it happens:

- Jake becomes a student of the Na'vi when he says to Neytiri, "If I'm like a child, then maybe you should teach me." She replies, "Some people cannot learn, you do not *see*." She is right. To his credit, Jake seems perhaps to know this and responds, "Then teach me how to see."[128] Neytiri says, with feeling, "No one can teach you how to see." And this too is true—unless the "teaching" is of an altogether different kind from what we typically think of as "teaching." The kind of education that comes from immersion and from within, from a passionate desire to learn, including crucially from those that one may have been told have nothing to teach. Complete respectful immersion in another, older/different life is required; or, failing that, immersion in another world entirely, where one is willing to learn . . . in our case, through the world-making properties of art and in particular of 3-D film (such as *Avatar*—and *Gravity*).

- And this is part of the significance of the remarkable scene towards the end of the film, in which, after saving his life, Neytiri cradles Jake in his (disabled) human form and says, "*I see you.*" She sees him, human, aware, whole and deeply loveable. Held in Neytiri's arms, he is now on the final stage of a journey towards going native. Jake transforms from being an American serviceman to become a neo-Native American. He makes the kind of transformational journey, the film intimates, that we all need metaphorically at least to go on if we are to save ourselves and our planet.

Does *Avatar* romanticize "the natives"? Well, it accurately *reflects* the genuinely ecological sensibility of some, perhaps most (*not* all) native peoples / native American "Indians" etc.[129] Take, for instance, the requirement to make no decision which shall harm the interests of the seventh generation, laid down by the Iroquois, still inspiring political thought today.[130] But the Na'vi are not depicted as saintly, not as homogenous, nor certainly as entirely peace-loving. Crucially, their tendency to righteous wrath and willingness to go to war are implicitly questioned in the poignant closing section of the film (discussed at length, below).

Does *Avatar* romanticize embodiment while also attacking and fearing the power that our bodies can have over us? Some critics claim that *Avatar* is prejudiced against the disabled.[131] Our hero Jake is wheelchair-bound, and though he is feisty and effective, nevertheless he wants to be able to walk, to run, (to fly!)[132] again. Well, it could surely be argued, so do many people with disabilities. But it is in any case a complete misunderstanding of the film to find here any prejudice whatsoever against the disabled, for the following four reasons:

1. *Avatar*'s disabled protagonist is a fighter with strong arms, able to take care of himself to a remarkable degree, despite the prejudice he encounters. "You never lose the attitude" of being a Marine, he says.
2. The prejudice that he meets is social, contingent. *Avatar* thus *endorses* a version of the "social model" of disability,[133] by making it starkly clear that Jake is deprived of the operation that would enable him to walk again simply because he lacks the money to pay for it. In the meantime it is the prejudice of his fellow humans and their failure to make an environment that works for him as a wheelchair-bound person which proves restrictive.

3. More importantly, not being able to walk is trivial compared with not having a clue as to how to survive, and indeed *not being able to breathe*. All *human beings*—all these "babies," one might say—are placed in the subject-position of the disabled on Pandora.[134] Thus a disabled human serves in this context *as a prototypical representative of the entire human race*. Jake's (contingent, simultaneously physical-and-social) disability stands in for *our* (contingent, simultaneously physical-and-social) disability in relation to the natural world, to Nature, that we tend to see as "Other"—and to a greenhouse atmosphere we cannot survive in. We *find* ourselves in a disabled character on-screen because we are *all* disabled (by Pandora). In other words, we all risk being disabled by/in nature *until* we find ourselves at home (t)here. We are all literally *en*-abled, by having an atmosphere we can breathe. *Avatar* radically turns around the social assumption of thinking of the disabled as somehow a perhaps-pitiable lesser version of the "able-bodied," and encourages/forces us to find ourselves *in* Jake. Far from being prejudiced against the disabled, the film's hero is an icon for humanity as a whole, requiring us all to really *think* about disability, to contemplate what it actually means, to experience the situation of the disabled vicariously, in order to be able to empathize with the disabled and know what life may be like for them, perhaps for the first time. I believe *the film does the very* opposite *of disparaging the disabled*.
4. Also, as mentioned above, the scene in which Neytiri saves Jake as he starts having to breathe the atmosphere of Pandora is a pretty devastating refutation of the "*Avatar* is prejudiced against the disabled" argument. The scene is deeply moving. Neytiri cries out, "Jake! My Jake!" when she finds his threatened human form close to death in the greenhouse atmosphere of Pandora (see [3], above).[135] Far from Neytiri being appalled when she finally sees Jake's *real* body, she is tender and loving. If viewers have been harbouring discriminatory thoughts towards Jake as a disabled hero, Neytiri shames those judgements and flushes them out into the open. When she says, "*I see you*," her words are the ultimate epistemological compliment, the truest expression of caring. Neytiri does not see a broken, disabled body; she sees Jake in his entirety, mind, body, spirit and soul, and this is the person she loves. This that she now cradles is the psychical and corporeal home, in a way the true form, of the man. (The best picture of the human soul is the human body.)[136] *Avatar facilitates* a deeply nurtured capacity for understanding and empathy, which leads to true acceptance and unconditional love.

It is in fact the soldiers who see Jake arriving on Pandora and Quaritch (the commander) who are prejudiced against Jake on the grounds of his disability. *Quaritch* says to him, "I'll see to it that you'll get your legs back. Your real legs."—but he makes this reward conditional on an act of treachery and complicity in genocide.

In fact: until Jake's lived consciousness changes, until he adopts ways that allow him to be part of the people and part of their ecosystem, then, regardless of the power of his new avatar body, in effect *he is* still *disabled.* The transformation one has to go through is far deeper than mere physical augmentation (e.g. being given new artificial legs). Rather, as Nietzsche would put it: it requires a radical self-overcoming. Jake will get new legs, he is promised, if he betrays the Na'vi. He holds off on getting the legs, because he starts not to (want to) betray the Na'vi. He is moving over, he is transforming: *not* for the purpose of getting artificial legs *nor* for getting new blue legs, but for what he is now coming to believe. For a genuine *purpose.* To fight for the living planet and its people.

Later, I will return to the question of disability in the film, in connection with a parallel between Jake and viewers, sat immobilized in a cinema seat. But first let us dwell further on the air of Pandora itself being death to humans, in terms of polluting our lungs and rapidly destroying us. Pandora may be a relatively unspoiled world—but it is a world where humans have (nevertheless) to face the consequences of an atmosphere that spells destruction for them (and that means: for us). Does this sound familiar? The hothouse atmosphere of Pandora *has very high levels of* CO_2.[137] It is unbreathable for humans.

Earth Systems Theory today argues that our atmosphere on Earth is breathable *because* of life. Because of "the oxygen revolution," the mass production of oxygen and reduction of carbon dioxide and other gases in past eras of our planet. *Avatar* highlights the profound sense in which (we tend to forget) life intertwines with the "environment," so much so that the "environment" has been radically reshaped by life. Is our reshaping of the atmosphere going to destabilize life on Earth? At present, it looks that way.

So, while Earth in the dystopian future imagined in *Avatar* has been utterly ravaged by humans and denuded of much wildlife and of many vital "ecosystem services," the utopia of Pandora carries with it a subtle but grim reminder of the most pressing current form of this ravaging destruction—namely, the awful damage that we are doing to our atmosphere. Will our atmosphere continue to be breathable? Will it always support human life, human civilization? Or will it in fact roast us alive, once we have exponentially destabilized it? This huge question looms large, giving us constant angst these days; and rightly so. *Avatar* does not let us forget it. If the Earth's ecosystems are destroyed through catastrophic climate change occurring as a result of changes that we are responsible for, it would make a grim kind of sense if we couldn't even breathe the atmosphere of a world that we tried to escape to.

However, we should note that the Earthlings in *Avatar* do not literally attempt to escape to Pandora—thus they are not in the customary sense of the word "settlers." The impression we have of the future Earth in the film is of a planet where human beings are systematically alienated from nature, and where nature has been despoiled yet more systematically than it has been in our time. In such a world, the predominant attitude towards nature will likely be one of fear and disgust, exactly the attitude that is indoctrinated into new Marine arrivals on Pandora. The way in which humans in the film treat Pandora, merely as a massive new mine for "natural resources," a "standing reserve" (in Heidegger's terminology), and *not* as a possible home, expresses profound alienation from nature. This is a central topic of the film, one which the film aims to midwife a birth away from.

Avatar draws you into Pandora and the native people who live there. In a way the film's narrative is cowboys and Indians all over again—*only this time, you're on the "Indians'" side.*[138] We vicariously join Jake in his progressive transformation into becoming one of them, while the invasion and destruction of this new frontier comes increasingly to seem an appalling thing. Incipiently: disaster triumphant.[139]

It is, though, a misunderstanding to dub *Avatar*, as many right-wing American commentators have, as "anti-American," for the following two reasons:

1. The "anti-American" charge is merely the flip side of an argument that has been made by "left-wing" cultural critics of the film, who say *Avatar* is tacitly racist or imperialist in having an American as the central character and seeing the whole situation (almost literally/entirely) through his eyes. Both charges are wrong. Jake Sully goes native. And takes us with him. The film *sets out*, and *makes available to experience*, exactly the kind of change that needs to occur in our world if we are to save the Earth and ourselves. Indigenous peoples alone are not in a position to save our common future. *We* need to actively learn from them, we need as it were to convert to their ideas and culture, and not to continually assume the reverse. Jake puts it thus: "There's nothing we have that they want." This may be true of many more indigenous people (especially the few remaining uncontacted tribes) and peasants than, narcissistically, we tend to suppose. *Avatar* disputes the gross falsification involved in the (very) idea of "(sustainable) development"[140]—the absurd notion that *we* are developed and that "they" need to become more like us. That idea needs to be stood on its head. The film helps us start to do that. We need to *do* what *Avatar* shows. But the *reason* why the (false) "left-wing" charge against *Avatar* is made refutes the accusation of "anti-Americanism." For it *is* precisely *Americans*[141] who play a crucial role in joining with Tsu'tey and others in leading the charge against the Marine attack. Especially, though not only, Jake and Trudy Chacón, the Latina rebel helicopter pilot. The truth is of course between—or rather, orthogonal to—the crude charge of racism on the one hand, versus that of anti-Americanism on the other; for the film does not generalize about Americans. It picks out a few individuals—including crucially

our narrator, Jake—as somehow being graced with the capacity to open to the transformation that needs to happen, as he becomes closer to the life that the Na'vi enjoy. It is not racist to try to save humankind by targeting your efforts directly on transforming the consciousness and practices of those currently responsible for most of the destruction. *It is common-sense.*[142]

2. Also, absolutely crucial are the film's "scientists," who are at heart anthropologists. Again, to a woman and man, Americans.

Admittedly, there risks being something unduly attractive to an academic about a film in which it is, most unusually, *academics* who save the world. Without Grace and Norm, there would be no opportunity for Jake to help lead the armed struggle against the Marines. Most importantly, Grace (her name no accident) plays a critical role in laying the groundwork for the psychical transition that Jake goes through, and that we go through with him. The two working closely together show how it needs more than a soldier / an individual to make the change. It needs thinking people open to truly learning from the other, to forge and *become* the change, together.

The key point in the process to reflect on is *what* these academics are. They are not mineralogists or physicists; they are biologists and *anthropologists*. They are seeking to learn about "the natives," but they are not classical anthropologists attempting to find out the strange ways of "the Other."[143] Under the striking leadership of Grace Augustine (Sigourney Weaver), they are genuinely seeking to learn from the other(s).[144] They do not regard the others' knowledge as mere superstition; they regard it as genuine candidate-knowledge. These anthropologists *have overcome the central prejudice of "social science,"* that this (or anything) is about "us" "scientifically" understanding "them," as our inferiors. Rather, what true anthropology ought to be about is finding out about ourselves through finding out about the other; finding out about the world through finding out about *them*; and finding out about them through genuinely *being with* them, *seeing* them, rather than objectifying them. As we (especially via Jake's journey) learn about the Na'vi's rituals, their ecology, their eating habits, their connectedness with creatures and with their Earth, we come to understand and appreciate and start to share in their wisdom. This *cannot be done* if one remains aloof and "superior." One has to learn with and from. To some extent, one has to become one of or with them.[145]

This process might be (and has been) termed "reverse anthropology."[146] As I say, this is anthropology/ethnography working as it ought to, removed from cultural imperialism and misplaced scientism.

As the Tsahik, Mo'at, says, the Na'vi will try to teach Jake, to teach the anthropologists and by default to teach us viewers: "[W]e will see if your insanity can be cured." So she instructs Neytiri to try to instruct him. Neytiri brings to him the indigenous practices and wisdom. And thus he brings it to us. She and he are the avatar(s) for us of that wisdom.

As she learns from the Na'vi and from studying Pandora's trees, Grace realizes, "The wealth of this world isn't in the ground, it's all around us." "Unobtainium," the name given to what the Earthlings are mining on Pandora, is in the end a blind, a MacGuffin: in this case, to superb Brechtian effect. The name "Unobtainium" helps make clear that the film is to be taken as a comment on or exploration of *our* world. *Avatar*, like *LOTR*, is a symbolical and healing psychical journey back to and through oneself, not an escapist fantasy.[147] It is quite risible that so few critics of the film have been able to see this, and have, in some cases, altogether laughably, taken the name "Unobtainium" to reflect a simple failure of imagination on the part of the film's creators. Why go to the trouble to create a *whole new Na'vi language*,[148] and fall at the first hurdle of coming up with a believable name for the mineral that the Earthlings are hunting? The name is a clue to the deliberately symbolic and allegorical nature of the work. It offers an invitation to the audience to reflect upon that—and, specifically, *to facilitate reflection upon what the Earthlings tend to miss about Pandora*. The critics who pan "Unobtainium" are failing to see the film in just the way that most of the Earthlings in the film fail to see Pandora! The real wealth of the world (as Grace spots) is not in its minerals, but in *life*, the networks of consciousness and communication and energy that interweave and interact. And these are uncapturable—unobtainable—by even the most sophisticated mining equipment. The real wealth of Pandora can only be "obtained" (the very word seems wrong) by one willing to get down and dirty *in* the world, to become part of it and its people. The true value of Earth cannot really be *obtained or owned* at all, only understood, participated in. For, as Thomas Berry has remarked, "The universe is a communion of subjects, not a collection of objects."[149] This notion of wholeness can only be appreciated by one *willing* to participate in and become *part* of Mother Earth's real wealth: this communion of humans, animals and plant life, under the sky.

This is the kind of learning that we need to bring to our world and that we desperately need to understand experientially, not just academically or scientifically.

Avatar: the very title of the film is a metaphor for experiential identification. But the key point is that the Na'vi people *are* people, centres of experience, immersed in a common life (as in a way we are, together, in the 3-D cinema).[150] Hence, one's/our avatar is engaged in a real-life, life-or-death struggle, initially with the "hostile" planet of Pandora—and, ultimately, with the American colonizers.

Here is how Vincent Gaine puts the matter:[151]

> While the avatar body is a form of augmentation, *Avatar* itself is riddled with these [with prosthetics], particularly visual augmentation, as Parker Selfridge and Colonel Quaritch primarily view Pandora on screens and through visual filters and barriers. The film places Jake Sully, in his avatar body, directly *within* the forest of Pandora rather

> than in command of it. Many shots present Jake as dwarfed by the jungle that both he and viewer can marvel at rather than control.[152] Visual augmentation is also unreliable: when a remote controlled viewer has its camera destroyed, its pilot proclaims in complete helplessness: "I'm blind." The instruments of the military personnel will not work in the Floating Mountains, and Jake comments that the soldiers must "fire line of sight"—use their eyes rather than devices.[153]

Of course, nestled within this quote is the paradox that it is only by Jake's "dropping" into a body that any of this is possible. The crucial opposition in the film, in the end, is between "dropping" into something, whether a mechanical prosthetic or an avatar, and *changing one's lived consciousness*. It seems only the latter actually yields potential enlightenment. As Joshua Clover sees, in calling the argument of the film

> that what we might call "vertical jacking" (as Jake into his vat-grown avatar and, by extension, a terminal operator into a Predator drone in Afghanistan, or you into *Second Life*) is bad, as it takes the fundamental form of domination: one extending its will into another. Conversely, "horizontal jacking" (with its at least purported sharing of wills) is just fine, indeed, it's "natural," even if conducted via technological means. And it is in fact a necessity in the face of unnatural domination, providing an alter-globalization.[154]

This is why the story has to end with Jake's enlightenment being achieved not by him remaining a drop-in, a dreamwalker, but by fully becoming one of "them," the Na'vi—and with his eyes looking out at us inviting us to take a similar transformative journey. (We'll come back to this one more time.)

Really seeing, and really being seen, as a dance of mutual acknowledgement and true vision. Arriving at the possibility of sharing, collectivizing of will. That is the challenge of the avatar—the challenge is to recognize "others" as real, to come to truly acknowledge them, as both different and the same.[155] This is of course *our task* too, as *viewers*. For what is the process of "becoming" one's avatar like? Is it like playing an intense, prolonged, character-based computer game? Possibly; but isn't it even more like *watching a film*, a film such as *Avatar* or *Blade Runner*? In a cinema, especially a 3-D cinema, one's involvement requires a bodily passivity reminiscent of what is involved in going into one of the virtual reality "coffins" in *Avatar*. This does *not* require the kind of frenetic physical activity involved in a Wii or computer game. This is of course how our Marine protagonist can have an avatar, even though he is paraplegic. While Jake "is" in his avatar body, his own body is as if immobile, as ours is while watching the film.

One might then suggest, building on what I set out already in the Introduction to this book, that *Avatar* is itself a metaphor for watching films; and especially for watching films like those explored in this book (including, most strikingly, *Avatar*). Unless you are *involved*, which includes being a *participant* in the social practice of acknowledging or otherwise, then you radically miss the point. The ultimate implication of *Avatar* (and of films like it) is that *the viewer's* actual *life* must be affected. Observing is not enough. Seeing is a prequel to doing/acting different.

You have to change your life.

What does our jarhead hero *do*, what is the crucial activity that Jake engages in on his gradual (staccato) conversion journey? *He makes a film . . .* a video diary. From early on in *Avatar*, the narrative *is* mostly taken (from) *the film that Jake himself is making, as he records his experience.* One might think of this as a metonym for the experience of making—or of really seeing—*Avatar*. A video diary records Jake's reluctant, surprising transformation into an eco-warrior. James Cameron created *Avatar*. So what are *we* going to create/do? This film about (making, and really seeing) films asks, I think, to be heard as a call for you to do something of a similar kind. To take the kind of action that really seeing the Earth, really thinking about and feeling and protecting, honouring our children to the seventh (and the 777th) generation, will require.

Avatar, as I have said, begins with a closed set of eyes, those of *our* avatar in the film. This suggests that our eyes are closed. It ends with the same eyes transformed into the eyes of a being who can now appreciate their embeddedness in the world and among others, the eyes of eco-sight, opening. This suggests that our eyes, as viewers, are now open. If we have really seen *Avatar* ("I see you"), it opens our eyes. It has opened our eyes. In this sense, we might say that the film *is* one slow gigantic movement of a pair of eyes opening, seeing as if for the first time.

Our eyes.

Through the eyes of our hero gradually being opened, we come to experience an "awakening." But the word "gradually" must be emphasized.[156] The process of therapeutic healing, the curing of Jake's all-too-human insanity—of his failure to acknowledge, to understand—comes painfully slowly, reluctantly. In fact the process is so slow that it almost comes altogether too late. Jake's awakening comes too late, too slow, to save Home Tree. (And ironically the video diary that our jarhead made is used as evidence against the values of the person he is gradually becoming.)

This gradual process is important because learning requires time. Time to make the journey and to wish that he and we were making it quicker. As Wittgenstein held:[157] in philosophy, a slow cure is all-important. Liberatory works of film need to proceed in the same way. To really take an audience with you, people have to become more than your *aud-ience*. They must not merely hear what you *say*: they must really see for themselves. Viewers have to say, as Jake does halfway through his transformation: "I don't know who I am anymore." We have to *work* through the therapeutic-liberatory transition that the film invites us into. This great work—the work of inner transformation,[158] that inevitably brings with it outer change too—cannot be rushed.

Our hero is exactly the kind of man who needs to take an Avatarian journey if our world is to be healed and/or saved. To be specific: Jake is a "middle American." He has been betrayed by his country, by large corporations, deprived of decent medical care. He gets healed by truly becoming himself,[159] through finding love and mutual care, in an ecologically viable setting. Through coming to live as his avatar does; through coming to be an avatar for Ey'wa; and, ultimately, through coming to *be* his (own) avatar. Jake eventually catches up with the being that walked ahead of him.

This vision of being and becoming is set against the closest counterparts to the avatars in the film, their "other": the giant robotic warrior suits used by humans without avatars to range out onto the surface of Pandora.[160] In the final confrontation between Colonel Quaritch on one side and Jake and Neytiri (who is riding the thanator)[161] on the other, avatar and Na'vi are ranged against one of these industrial fighters. The contrast could not be more striking. The avatar brings one into closer-than-close contact with the environment, with nature. One lives naturally, whereas the oversized metal fighting suits seal men off from nature (possibly even their own nature) and set them on a crusade to "overpower" the environment. This is the opposition: the possibility for transformation and a finding of a harmony with an unsentimentalized, red in tooth and claw[162] nature, on the one hand, versus military-industrial othering from nature, on the other. Avatars/people/animals—versus machines. This othering, the mechanized *distance* from nature and from reality, contrasts strikingly with the achievements of Grace's anthropologists.

When the ultimate military enemy of nature, the villain of the piece, Marine Colonel Quaritch is finally killed, it is by our heroine, Neytiri. She transfixes him with two arrows. As the second arrow hits, we are given a point-of-view shot *from Quaritch's perspective*. In other words: we experience his dying along with him. This suggests that the kind of American Quaritch represents and that exists in most or all of us, also *has to die. We have to die in order to be reborn.* The film's "message" at moments like this, i.e. the therapeutic (healing) journey that it encourages viewers to take, is deeply challenging.

Avatar offers the challenge more than once. At the start of the film we learn that Jake's twin brother—a doppelganger for our hero—is killed on an Earth that has literally lost its sense(s). A powerful point-of-view shot places us for a little while inside his brother's coffin. (The coffin stands as a proxy for the body-chambers that later transform humans such as Jake into avatars.) We hear Jake say, "One life ends. Another begins." Indeed; to be born again, first you have to be willing to die. As a Marine, Jake is accustomed to sacrifice, although at this stage of the narrative he is unaware of what his "death and rebirth" will cost him personally. All his former allegiances to military life will be challenged and transformed, in order that he may be free to fight the actual good fight. At the end of the film, when he makes his final video log before, he hopes, becoming a Na'vi forever, Jake says: "It's my birthday, after all." This echoes his earlier remark which, in turn, echoes traditional "rites of passage" practices in indigenous tribes that unfortunately we have become somewhat remote from today: "The Na'vi say that every person is born twice. The second time, is when you earn your place among the people forever."

You have in *this* way to be "like a baby," ready to let the Earth, creatures, indigenous peoples—and new mythic works such as *Avatar*—re-shape and educate you. Again, this is "reverse anthropology." As one of Grace's team puts it: "We've got to get in the habit of documenting everything, you know, *what we see, what we feel*, it's all part of the science." Think about that statement. This is in fact a radical re-conceptualization of what (human) "science" is.

And why do I scare-quote the word "message" in the paragraphs above? Because when one really understands films like *Lord of the Rings* and *Avatar*, they don't have generalized messages as such. They are not disguised bits of propaganda. Because they *essentially* involve the viewer. They guide the viewer on a proposed "journey," yes—but the journey is psychically individual, as well as partly collective. (As noted in the Introduction to this book, I think it important that we see these films *in cinemas*—I return to this point, below.) The specificities of each person's journey will be different, of course; and indeed, one may refuse altogether to take the journey (as many critics have done). Part of my account of these films is inevitably autobiographical. (I am allegorizing *my* reading/viewing of these films. The "message" that I speak of is thus the message *for/from me*; and everyone, each person, must in this way speak for themselves.)

Films such as *Lord of the Rings* and *Avatar* do not strictly speaking make *arguments*, nor exactly do they have "morals." Rather, they offer what Wittgenstein, Buddhism and emancipatory traditions alike call *liberation*. This is philosophy not as theory nor as quasi-factive impersonal claim, but as a *process* one must work through. It is different from the idea of philosophy to which we are accustomed; it sits ill with the idolatry of science which lies at the heart of our civilization.[163] So much the worse for that idolatry. It is idolatry of science and the taking of technology as a "neutral" tool that has got us into the proto-catastrophic mess we are in. *Avatar* dramatizes and extends the "logic" of this. Thus we should *expect* that a non-scientistic vein of philosophy, such as Wittgenstein offers, is appropriate to help us understand how to extricate ourselves from the mess. Fortunately our expectation is not disappointed. The film, like Wittgenstein's writing, is designed to *heal*. But healing, genuine healing of one's mind, one's body-self, and of one's world, is properly an art, not a science, and is throughout processual. "Healing," moreover, etymologically means "to make whole"—the latter being something I've emphasized throughout this chapter.

Thus *Lord of the Rings* and *Avatar* are *not* escapist (unlike, say, video games, where one also has an "avatar"). The films thematize an *illusion* of escape: that's what it is, to think of putting on the One Ring; and it's also what Jakes aspires to "as" his avatar, until he passes through the painful process of enlightenment. When one has overcome that temptation of escape, these stories return one to oneself and the world, perhaps ready to know it—to feel it—for the first time.

This is what I see in these films. But I do believe it is, to a greater or lesser extent, consciously or unconsciously, what many millions of others see too. This would help explain the films' vast popularity. *Lord of the Rings* has, seemingly, multiple fairly obvious flaws, as delineated earlier; *Avatar* could easily be seen as a predictable and just very shiny exercise in "cheese," or as a rant. Many critics have responded to *Avatar* either from "the Left," with cynicism and a knowing superiority to such alleged sentimentalism, romanticism and superficiality, or with allegations that the film is tacitly racist against indigenous peoples, against the disabled etc. as discussed above; or from "the Right," with anger against the attack within the film on cultural norms, on American militarism etc. If anything, I think the critics from "the Right" are closer to the truth, despite themselves. The film *is* shocking, in the extent to which, when one experiences it closely—when one experiences for instance the arrow transfixing and killing one's American/military/racist/speciesist self so that the world can be saved—the journey it proposes takes one very far from one's comfort zone. I think the reason why the film has been found by so many to be emotionally compelling—as emotionally compelling as the Na'vi themselves are in their emotional healthiness and expressiveness—IS something like the line of understanding of the film that I offer. People find *Avatar* compelling *because* of the journey it takes them on, *because* of the assumptions the film puts into question, *because* of the way it speaks to our condition as being alienated from our planetary home and from each other. And this is why *Avatar* was banned in China.[164] This is why the film inspired colourful protests against the apartheid wall in Palestine;[165] and why the Dongria tribal people have invoked *Avatar* in their successful struggle against extreme extractivism;[166] and also how and why it is inspiring the activist work of the Radical Anthropology Group.[167]

The same is true of *Lord of the Rings*; the drastic plot flaws and unbelievable nature of the narrative end up if anything closer to *pluses* than minuses. They provide gentle, tacit "alienation effects" roughly in Brecht's sense of the word; they *enhance* the experience of questing that the viewer vicariously has. The psychological journey that one is taken on, into oneself, into one's courage (or lack of it), into one's faith in oneself and in others, and in what Aragorn calls "this good Earth."

I will return to how *Avatar* is *unbelievable*, larger than life. But, given the argument of the previous paragraph, we should note why the Na'vi are literally larger than life. They dwarf us humans and in doing so make it easier for us to think about them as people, for all their so-called "primitivity." The Na'vi figure as our new parents. Again, this is partly alluded to in the scene where Neytiri cradles Jake's broken human body and responds lovingly without patronizing him. This is part of a journey to maturity that she sees him through and joins him in. Neytiri gradually moves from being Jake's mother figure and teacher to being able to love him as her equal, a point she arrived at prematurely earlier in the film, to great pain and cost. Finally, Neytiri sees the beauty of Jake's form, of *her man* in his totality; healed and whole. (*Together*, they are, as one might put it, doubly complete.)

The Na'vi representing "new parents" is a clever *bouleverser* of the standard stereotype of "primitive" peoples as being like children. Similarly, and as discussed further, there is an obvious connection with Jake referring to human activity on Earth as meaning that humans have "killed their Mother"—and so are badly in need of new parents. All the same, it might be claimed that there is something odd and Oedipal about the scene I keep returning to, in which Neytiri cradles Jake. The visual "iconography," one might say, is of mother and child. But one has to place the scene in the context of their developing relationship, a relationship that is complicated because of Jake's witting and unwitting role in the betrayal of the Na'vi people, which led to the destruction of Home Tree. He wasn't mature enough, it turns out, to be Neytiri's equal, *until* the point when he finally "goes native" and turns and fights against Quaritch, the "father figure" who repeatedly addresses him as "Son." By the end of the film, Jake no longer needs mothering by his lover. He has truly learnt—and so, perhaps, have we viewers—that Ey'wa (Gaia/Mother/Sister/Daughter-Earth) must be respected not raped, that She is alive in all her fruitfulness and glory; and that future generations are collectively Her children, that it is *they* that need virtual "parenting" by all of us, and that we have to be mature enough to ensure this happens.

That's why Neytiri can say "I see you" to Jake, in his human form, *transcending* the physical differences between them.

And what of the planet that is Ey'wa's body, the Na'vi's mother? Why is the planet (the moon, to be precise) called "Pandora"? Perhaps because *Avatar* offers us *hope*. In the original myth of Pandora's box, its opening initially releases poison and awfulness;[168] but a gorgeous, vital sliver of a silver lining then comes to light and this very discovery constitutes hope. Pandora features a host of "natural evils" (as part of its nature); moreover it unleashes the worst in humanity in terms of grabbing at its "natural resources."[169] The mega-machine, the juggernaut of industrial-growthist destruction, triumphs over the Na'vi rebels and their few human allies. But yet hope arises from the planet itself.

The hope is ultimately vested in the viewer. The hope is that, with the wisdom of what we have learnt from the film, we can find a route to stopping the juggernaut of "progress" and profit before it is too late, before our planet is wrecked. We can prevent the opening plot device of *Avatar*—where the home of human civilization is a deeply unjust place—that depicts Mother Earth dying with her ecosystems terminally ruined—from becoming true. In this sense *Avatar* aims to be a self-defeating "prophecy." The film is a warning, we might say, from the future, as *Lord of the Rings* is from the past.

Just as Sauron (and the One Ring) is an icon of the deadly fantasy of absolute safety through absolute power, so more straightforwardly is the human world (the humanism, in the worst sense of that term)[170] of *Avatar*, which *is* the hegemonic tendencies of our world today projected forward into the future. A world in which we imagine, absurdly, that anything is possible for humans; in which we presume our technological mastery of our environment is absolute. We fantasize that we can control Nature herself, that we can mine Mother Earth and rape and pillage/pollute her as much as we wish without her being ruined—and without the murderous intent to harm coming back to bite us. We fantasize that we can be safe from our projection of a disgusting, fearful, threatening nature by endlessly increasing our powers of prediction and control over her. But the truth is that such a project endlessly increases our exposure to risk and uncertainty, and ultimately such hubris brings nemesis. Ecological disasters and ecological cataclysm. That, Jake tells us, is the fate of Mother Earth . . . unless we learn the lessons of *Avatar*.

And now perhaps Ey'wa's revenge at the end of the film can be read in a slightly different light: as a metaphorical picture of what Lovelock calls "the revenge of Gaia," which will be wreaked upon our species unless we manage to learn to change our ways and learn fast.

The hope unleashed by the opening of Pandora's box is vested, in the real world, in *us*. This is tacitly true of the original Greek tale—hope is personified, at the bottom of the box; but hope is only actually real if it is individually and socially made real (i.e. by real persons). The hope offered by Pandora, by *Avatar*, is that the hope is in *us*. That we can all be part of fighting, struggling, intelligently and successfully, to save ourselves from the future gestured at in the backstory of *Avatar*. The hope is slim because it requires an unprecedented revolution, going far beyond *anything* that is mainstream in our current politics. The hope rests then on the kind of faith, faith beyond any merely realistic or "evidence-based" hope, that is cultivated in *Lord of the Rings*. Faith in ourselves and each other and our Earth, even when—in fact, especially when[171]—all *reasons* for such hope have run out. (As in the Warsaw ghetto uprising; or in the last moments at Helm's Deep.)

This is the kind of philosophy we need for the 21st century. The groundwork was created, as I noted above already in considering *LOTR*, by Pascal and (better) by Kierkegaard and (best) by William James. If we look only to reason and to the facts to give us hope, then hope/we will die. Obviously we need to reason and we need science and we need to stay in touch with the facts; but above and beyond that we need to trust, to have faith, to believe even when reason says there is no reasonable hope and no reasonable doubt about the fate that lies in store for us. In Pascal's terms:[172] if we do not wager, if we do not act as if there is a chance that we can save ourselves, if we refuse to take the risk of holding out hope that we may be able to save ourselves, then certainly we will fail to save ourselves. In Kierkegaard terms (from *Fear and Trembling*): faith is most truly faith, *when* it seems absurd. In James's terms: we have the right (the "will") to believe certain kinds of things—such as to have confidence in our collective goodness, in humanity having the capacity to achieve balance and ecological sanity—even without evidence;[173] but, *without* the will so to believe, we will in such cases lose our only chance to achieve the fruits of such belief.[174] The challenge is to embrace this vital creative truth without our doing so degenerating into being one more version of humanist cognitive hubris, denial, over-reachingness.

Faith makes possible what for reason alone is impossible. This is a central teaching of *Lord of the Rings*, as I have sought to show. (Recall Aragorn's "Lean forward" moment, and all it symbolizes.) It is also implicit in *Avatar*. It is gestured at every time Jake's avatar makes an improbable leap, or wins communion with one of Pandora's creatures of the air. "Con-fidence." With faith.

Both *Avatar* and *The Lord of the Rings* examine the attractions of the tendency to retreat. To give up hope. The very temptations analysed by *Lord of the Rings* and *Avatar* constitute the main reason why so many people are unprepared to embrace the challenges contained in the narratives and prefer instead to back away; to ignore or stand aloof from them.

These films aim to overcome such hopelessness, such giving up on humanity. They aim, therefore, first, to understand it (i.e. to understand why hopelessness and cynicism are so attractive). It is not surprising if a major reaction is incomprehension of and more-or-less politically motivated resistance (which is also psychologically motivated, i.e. as a defence mechanism). *Lord of the Rings* and *Avatar* invite viewers *to dare to hope*, which is a difficult invitation to accept, especially when there is so little basis for optimism. Yet it is precisely then (i.e. now) that we need such daring invitations, in order to start to make possible what the cool rational mind considers absurd, hopeless. *Avatar* and *Lord of the Rings* invite us to take the risk of hoping, of not giving up faith in ourselves and life. Much of the resistance to the films is, in my view, simply disguised hopelessness. *The critics who scorn are exactly those most in need of their "therapy."* The resistance to *Avatar* is exactly what *Avatar* is *about*. If the film had not provoked the kind of negative reactions that it has, in fact, one could be fairly confident that it was not as great or as needful a film for our time as it is.

This is exactly the kind of thing that Wittgenstein meant when he said that philosophical problems are ultimately problems of the will, not of the intellect. What we as a species need is not to become even cleverer; what we need is to want enough to get well, to heal, to sort ourselves out. We need to want badly enough—we need to *will*—the saving of our common future, by making real change, on personal and worldly levels. We need to treat our own inclinations to resist a film such as *Avatar* not as intuitions to build on, but as inclinations that require philosophical/liberatory/therapeutic treatment. Like Jake, we need to lose our attachment-based fear and be brave enough to transform our ego. We do not have power over Mother Earth. If we try to control her, we will encounter the kind of nemesis that climate chaos already seems to be bringing. What is called for is a new respect for the bounty and beauty of the Earth if we are to restore Nature's and our own delicate balance. A new willingness to live within natural limits.

Avatar, more explicitly than *Lord of the Rings*, and perhaps even more poignantly, issues to us a *call*. As I have implied, the call is not simply a call to arms. It is crucial to understand the profound lesson that the battle which ends the film teaches.

The struggle against all odds to fight back against the Marines with military means *does not succeed*. Its failure is captured most profoundly in magnificently mournful music—including tremulous solo wails of pain—accompanying the military *defeat* of the attack on the Marines (as earlier, following the destruction of Home Tree). This is amplified or focused by a terrible image of one of the Pandoran horses that had borne a Na'vi warrior (now presumably dead) *ablaze*, and in the consequent despair of Neytiri as she prepares to throw her life away attacking the Marine lines.

Against the crucial charge that *Avatar* glamourizes violence, romanticizes the anti-colonial struggle in an apolitical way, and offers a fairy-tale ending, I believe that we can reply in the negative. Yes, Ey'wa comes to the rescue. But we know that Gaia will not. Much like Frodo's quest in *Lord of the Rings*, and like the (diversionary) military struggle called "The War of the Ring," the crucial thing to remember is this: that, on its own terms, the rebellion of the Na'vi against the colonizing Marines *fails*. It is a glorious example of heroic virtues in action, but it does not succeed. This is the great poignancy of the battle. We, as viewers, so desperately (possibly even to our own surprise) want the American forces to be killed by the rebels and the Na'vi; we so desperately want the Na'vi to win. *But they don't*. Only the *deus ex machina* of Ey'wa yields victory. And this cannot be hoped for.

We have to create it ourselves. We collectively have to embody this deity; *we* have to become and *be* (as, or truly with) Ey'wa.

Avatar powerfully motivates a (temporary) hatred of the American soldiers who continue to obey orders.[175] This, of course, has contributed to right-wing American criticism of the film. It is remarkable how much, in the massive final battle, one is willing the Na'vi to succeed in beating/killing their colonialist attackers. Viewers are familiar with films in which humans counter alien attacks on Earth; but the paradigm is inverted in *Avatar*, as the aliens try to beat off attacks from *us*; and we gradually, perhaps painfully, come to adopt their point of view rather than "our own." Part of the "therapeutic" work of the film is to motivate and enable this striking, surprising desire. For example, the point-of-view shot through which we see the hatred on Neytiri's face as she unleashes the arrow that kills one's Quaritch self.[176] *But*, this is only a movement, a moment in the film; one does not end there. For when one has seen the film, one knows that the violent rebellion of the Na'vi, just and dignified though it is, and without alternative, failed. The *deus ex machina* that secures success and a happy ending is a *deus ex ey'wa*, or a *deus ex gaia*; in short, a *deus ex deus*. But we know that a g/God or g/Goddess alone is not going to save us. We have to do it ourselves, every man and every woman, together. Collectively, we have to find a way to secure our future that works. We need to persuade millions of people to actually take up a non-violent way forward; for the enemy, in consumer society, is in a certain sense *us*. *We* are complicit in devastating our planet, devastating our future. The call is to all of us, and we need to respond intelligently and organize accordingly, persuading others to take the journey with us towards resilience and restoration, if we are to succeed/survive, let alone thrive.

Like *Lord of the Rings*, then, *Avatar* is *not* a pro-violence film. It does not call for violent ecological civil war. If you take the military on "headfirst," you will likely lose. The call is for viewers to use their heads. In this sense we are collectively called on to proceed headfirst, intelligently.

To the ultimate objection that is made against *Avatar*, that the film's ending is unbelievable, and is thereby an unacceptable romanticization of hope, we can in fact accept what factually motivates the objection, but refuse the claim that it is an *objection*. The unlikely ending is a great *strength*, not a weakness![177] Like some of the other films examined in this volume, *Avatar* does jujitsu on the audience because even its "weak" points are actually strengths. In other words: this grand narrative "deficiency" is in fact the cleverest of twists, a great achievement, for it suggests what in a sense *ought* to happen. The planetary ecosystem, Gaia and all creatures of our world ought to rise up in horror and fury against our current ecological challenges. But: Gaia will not strike back.[178] The end of *Avatar* has to be unbelievable, so that we can become clear about the difference . . . between the "fairy tale" and our actual situation. (In this sense, *Avatar*'s ending functions like another of what Wittgenstein calls "object[s] of comparison" [see *PI* 130–2].) Only with such clarity can we move forward and start to do what is necessary to prevent us from killing our Mother (Earth).

This is of course why *Avatar*, like *Lord of the Rings* but still more so, is a film that actually might help save the world. The central struggle of our time is to change the practices of billions of people on the planet. How can such numbers be reached if not through the most successful film(s) ever made (and crucially, in the case of *Avatar*, their coming sequels)? Through a film that issues a call, that midwifes a change in conscience and in consciousness. A film that requires an emotional, thoughtful and practical response, one which suggests that, outside of a fairy tale, there is a route that must be found and (with will) *can* be found to ensure that the future it depicts for us/Earth does not materialize.

Jake tells us: "See, the world we come from, there's no green there. They killed their Mother. And they're gonna do the same here."

The call is to become eco-warriors who can win; and that means no unnecessary heroics. This requires bold but carefully judged *political* (and cultural) strategy, not outright attack. It requires the winning of "the climate war", but also a more sustained addressing of the general ecological crisis of which man-made climate change is just part (the most short-term pressing phenomenon). The climatic canary in the coalmine is telling us our way of living will change—either by our will, or by way of catastrophe and collapse. Our world requires the kind of ecopsychological transformation that *Avatar* aims to birth in us.

The call is to open (y)our eyes and act, before it is too late, to save this beautiful planet. We have to learn to think for the future, to think collectively, in fact to think *as an ecosystem*.[179]

The most powerful moment of hope in *Avatar* is the very last shot, which mirrors the opening of the film as discussed earlier: Jake's eyes flick wide open; now he has transmigrated fully into his avatar body. He has become his own avatar. It is worth dwelling on this. What do we, as viewers, see? What are we seeing when Jake's avatar eyes flick open?

I already wrote, in the Introduction to this book, of how this may be read as a look of love. Jake's eyes, as they open, will see the face of Neytiri, who looks to be leaning over him. Eyes wide open is how we might describe *her* point of view, which we perhaps now temporarily occupy. This is interesting; in the very final shot of the film, we see, arguably, *from the point of view of the central female Na'vi character*. We see, as Jake does, through the eyes of the beloved, a daughter of the great Mother. This is subtle, as easy to miss as the fact that *we* die as Quaritch, when he is transfixed by Neytiri's second arrow. We witness Jake's final transformation into one of the natives, the Na'vi; but we do so from a subject-position which is *already* that of one of "them." The shot perhaps implies we may have, in some sense, made this transformation (for) ourselves.[180] Our avatar may already be ahead, waiting to meet us, if we (are) AWAKE.

We can therefore add that this is perhaps a look of enlightenment. Jake has left his human form behind; he finally fully knows what it is to see like a Na'vi, in a Na'vi body, from a Na'vi perspective (rather than as an avatar).

Most important of all, *he looks directly at us*. Viewers receive him directly. He looks us straight *in the eye*—which serves to complete the call to action. It is as if Jake is asking, "*What are* you *going to do now?*" The eyes address us directly and require a response because Jake, our hero-catalyst, activates a *new point of view*: one that we might now think of as Neytiri's, his *and* (y)ours, simultaneously. In a certain sense the film is not over until we complete it—through our own transformation and our own action—for, if we do not act differently, we have not really been transformed.[181]

Because, let me close this chapter by noting something strikingly unusual about the hero's journey in *Avatar*. The normal structure of the journey is that the hero returns, enlightened. This occurs very obviously in *Lord of the Rings*, in *2001: A Space Odyssey*, in *Gravity*. In a sense it occurs in *Apocalypto*: Jaguar Paw gets back to his village just before the end of the film, though it is unclear that he is enlightened; rather, as I explained in Chapter 1, it is we who gain the enlightenment, through the final twist when he carries on beyond his village to the sea. In *Waltz With Bashir*, the situation is again slightly more complex: the protagonist apparently "returns" to Sabra and Shatila, but we can take it that this is an enlightened vision, not a psychotic break. In *Solaris*, the return is fake; it occurs only at the level of delusion. And that is the risk, if the hero remains in the "special" world (Solaris, Pandora).

And yet it seems plain that (one feels that) *Avatar*, unlike *Solaris* (or *Last Year in Marienbad*, in which there is no return, but only an endless stuckness), *does* somehow involve a completed hero's journey.

How can one square the circle here? How are the lessons brought back from the "special world" to the "ordinary world"?[182]

The answer is already implicit in my argument above—and in Chapters 1, 2, 3, 4 and 5. It is that we, the audience, have a position structurally analogous to Jake's. Our ordinary world has key features in common with his; it is alarmingly lined up to end up being like his. He has an avatar; and so do we (i.e. him).

He does not return from the special world; he enters into it more definitively. In a way, that's what we have to do to; that's the learning we bring back. And: we bring it back *when we return*. When we take off the 3-D glasses and leave the cinema.

We have to supply the "elixir"; we have to come back from "the special world" and make *this* world *special*. The call is not for us to become depressed: it is, as I have emphasized throughout, for us to *feel* the sorrow, the beauty and the need of reality. To become *aware*. And then we won't retreat from our agency any longer.[183]

Notes

1. I wonder whether the vast popularity of *Lord of the Rings* today can be explained as other than it being a particularly good epical swashbuckling battle-adventure yarn in a (technically marvellous) depiction of an alternative world. I think it can. And, as will emerge, I think the popularity *cannot* actually be explained through its being (e.g.) a piece of particularly well-plotted swash and buckle—partly because, as I shall explain, it simply is not.
2. There are frequent differences of nuance and of detail between Tolkien's epic and Jackson's. Where they differ, I almost invariably stay closer to Jackson's version. It more consistently yields I think the profound explorations and truths with which this essay is concerned. For all its apparent swashbuckling semi-Hollywood-ness (see n.1, above), the Jackson trilogy usually enforces on us more of the troubling psychological journey which is to me the nub of *Lord of the Rings*. Indeed, one might venture that it is the overcoming of the swash and buckle, of the impulse to read the fighting literally, that is one of the film's deepest meanings. One learns most deeply from *Lord of the Rings* if one is successfully taken in for a while at least by its seeming message of "emancipation through violent vanquishing of Evil" (cf. n.4). As in Wittgenstein's work (e.g. PI 119): if one was never inclined to judge a nonsense, then one would miss out on the learning—the "know-how"—that comes with coming to (and learning how to) overcome that inclination. As one can do especially well when one has a year or more to reflect, as was the case between the issuance of the *LOTR* films.
3. To use the wonderful term coined by Louis Sass in his book of the same name (Ithaca: Cornell, 1994).

4. Examples of films that do this are legion: they include some of the most famous films about madness, such as *The Snake Pit* or *Raising Cain*. *A Beautiful Mind* is better, because it is less obvious, more inhabitative, in key sections of it. Films such as *Fight Club* and *Memento* (see n.11) come close to fully pulling off the trick of successfully inhabiting madnesses throughout their length. But even they are, in my opinion, inferior to *The Lord of the Rings* in this regard, due to the irony that they STILL seem too much to be . . . about madness.

   On this, see also Walker Percy's The Moviegoer, whose epigraph, deeply relevant to the present chapter (including to how a deep film about madness has to not seem to be too much about madness), is this, from Kierkegaard: "The exact character of despair is this: That it is does not know that it is despair."
5. I refer here to The Divided Self (Harmondsworth: Penguin, 1965), not to Laing's more "constructivist," more anti-psychiatric, and generally less sophisticated later works. Those later works are I think right that the person in psychosis can and should never be given up on (see my argument concerning *The Two Towers*, especially); but they oversimplify and sometimes falsify the origin and nature of the corrosive self-defeating mechanisms that foster and maintain the psychosis in the first place, and thereby they tend to romanticize psychopathology and to exaggerate the degree to which an insane society can be blamed for individuals' "insanity."
6. Compare Wittgenstein's remarks about God in the Lectures and Conversations on Aesthetics, Psychology and Religious Belief, edited by C. Barrett (Berkeley: University of California Press, 1970).
7. Madness and Modernism (Cambridge, MA: Harvard University Press, 1992), p. 352.
8. See especially Sass's Paradoxes of Delusion, p. 71f.
9. The first monstrous words that Frodo very loudly yet very vaguely hears from Sauron, while wearing the Ring, are, "You cannot hide. I see you. There is no hiding from me, in the void. Death!"
10. To avoid a misunderstanding: I am not simply urging that one should always "stay" in the consensual world. No; sometimes withdrawal, meditation, questing are a necessary PART of ordinary life. Indeed, part of the point of this chapter is to unveil clearly an aspect of *Lord of the Rings* indicating that the viewer's experience of it is itself more like a meditation or at least a contemplative quest than it perhaps was before; and I think this a good thing.
11. Somewhat similarly: we naturally empathize with Leonard Shelby in the very fine philosophical thriller, *Memento*. It comes as a genuinely shocking aspect-shift when we find at the "end" of the film that even he is very far from being a "goodie." Furthermore, the philosophical pay-off of *Memento* comes in part from working back and forth between Shelby and ourselves to see how he is both more different from us and more similar to us than we had thought possible. As with Frodo (and Gollum, and Sauron, . . . etc.).

    For more on this, see my paper written jointly with Phil Hutchinson, "Memento: A philosophical investigation," in my Film as Philosophy (London: Palgrave, 2005).
12. And here we find a nice irony. Our spectatorship perhaps demands great and greater overwhelm, worse and worst odds, as the films go on. Is the audience to *Lord of the Rings* caught up—unconsciously? . . . deliberately?—by the films into an addictive spiral of vicarious fear, a spiral of power and pleasure which mirrors the very dynamic that I am urging the films expose? I suspect that it is, rather as *Natural Born Killers* is caught up in the very spectacle of violence-voyeurism that it satirically (and brilliantly) denounces. Perhaps this is in part because *Lord of the Rings*, after Tolkien, is itself split between

its heroes, between for instance Frodo's non-violent dissolution of power and Aragorn et al.'s violent possession of it.

But cf. also my discussion of how we should construe Sauron's (and Saruman's) forces and their very existence (or otherwise).

13. As suggested by Mark Eddy Smith, on pp. 108–9 of his Tolkien's Ordinary Virtues: Exploring the Spiritual Themes of The Lord of the Rings (Downers Grove: InterVarsity, 2002)
14. In this metaphor, I take it, the Ring itself and the Eye, whose pupil is formed by the fire around it, come together.
15. This is a kind of internalization of Sauron's lidless tireless Eye—Frodo feels completely and constantly Seen, and thus is constantly aware of the presence of the alleged Seer. It is this being/feeling seen that is perhaps worse than the seeing (of hallucinations). (One might compare here also *Donnie Darko*.)
16. P. 67, David Loy, The World Is Made of Stories (Somerville, MA: Wisdom, 2010).
17. Oppression and Liberty (London: Routledge, 2013), p. 62.
18. And eventually like Frodo. With Sauron, it is explicated in Gandalf's words, "They are one: the Ring and the Dark Lord."
19. The two combine in the discussion of Gollum undertaken in "A precious case from Middle Earth," by Sampson Bashir et al., *British Medical Journal* 329 (18 Dec. 2004), 1435–36, wherein the diagnosis of Gollum as suffering from "schizoid personality disorder" is preferred to a severer diagnosis of schizophrenia. However, while I certainly agree that there is much to be said for this diagnosis, I believe that the authors have not considered the essential paradox of the depiction of severe psychopathology: that a phenomenological or interpretive or descriptive work about (for instance) schizophrenia must not seem too decisively or definitely to be about schizophrenia, on pain of failing to capture the full horror of endless uncertainty about what is really happening and what is one's own delusion (cf. n.2 and n.4, above). The authors are too sure of a pre-Sassian "medical model" concept of delusion as "false, unshakeable beliefs"; this misses both the extreme reasonableness of the path to delusion and the subtle and ambiguous character of delusions themselves. These are "made perspicuous" by the allegorical character of Lord of the Rings, i.e. by its not seeming to be primarily about "madness."

   Compare also these important remarks from Terence des Pres's The Survivor: An Anatomy of Life in the Death Camps (Oxford: OUP, 1976), p. 81f.: "Speaking of his own camp experience, [Bruno] Bettelheim observes that 'right from the beginning I became convinced that these dreadful and degrading experiences were somehow not happening to "me" as a subject, but only to "me" as object.'" This has probably rightly been described by Viktor Frankl as "a necessary mechanism of self-defence" (Man's Search for Meaning [New York: Washington Square, 1984 (revised and updated version; original, 1946)]), p. 39; I have sketched above how it is also in itself a very perilous one. Des Pres goes on: "For [the majority of survivors], entry into the camp world was characterized by an overriding sense of nightmare and unreality—two words which appear constantly when survivors refer to their first days and weeks: 'All around us were screams, death, smoking chimneys making the air black and heavy with soot and the smell of burning bodies. It was just like a nightmare and it took weeks and weeks before I could really believe this was happening.'" The words are strikingly reminiscent of words commonly used by schizophrenics, who typically feel their life-world to be unbelievable, unreal, rather than simply being thrown into it as if in a Dionysian trance—see Sass's work for examples.

   Something that fits glove-like with my analysis in the present essay is that a number of survivors attribute their disbelief, plausibly enough, to "faith in humanity" (Des Pres, p. 83): to a felt impossibility in believing that all this could possibly be being engineered by human beings, by real natural non-monsters, "fellow men." The task is perhaps one of retaining faith in humanity, faith in life, once one's reasons for it, evidence for it, have gone. That task may well be pragmatically impossible, though, if one fails to find any society—any new "evidence" for faith in humanity—in one's fellow camp inmates. See on this the discussion of (Des Pres and) Levi, supra to n.22 & n.27.

20. Recall Gandalf's interpretation of Gollum's mindset and significance, when first he describes Gollum's encounter with Bilbo to Frodo: "Even Gollum was not wholly ruined. . . . There was a little corner of his mind that was still his own, and light came through it, as through a chink in the dark. . . . It was actually pleasant [for him], I think, to hear a kindly voice again, bringing up memories of wind, and trees, and sun on the grass, and such forgotten things. // But that, of course, would only make the evil part of him angrier in the end—unless it could be conquered. Unless it could be cured" (*The Fellowship of the Ring* (London: George Allen & Unwin (Anchor Press), 1981 (1954), p. 83).
21. And bearing in mind that Tolkien wrote *LOTR* during and in the aftermath of World War II.
22. Levi, *The Drowned and the Saved* (London: Abacus, 1989), pp. 63–64. An analogy would be: *Lord of the Rings* should be written by Gollum (or even by one of the Ring-wraiths), not by Frodo. Only, that might beg the question over whether even Gollum has touched bottom. (And of course, and this is in part Levi's point: "muselmann-ness" is the absence of a work. There is no story, from the Ring-wraith's "point of view." I expand on this in the text as this chapter develops; see also p. 88f of Des Pres's *The Survivor: An Anatomy of Life in the Death Camps*. Cf. also the early Foucault on madness as the absence of a work.)
23. This may appear to be the state of Saul, at the beginning of *Son of Saul*. The film is about how he recovers from that state.
24. London: Abacus, 1979, pp. 155–156. *Even if we return home*; the fear can be expressed, in mythic terms, as the fear of becoming constitutively unable to complete any "hero's journey".
25. My point here runs closely parallel to the point I made in Chapter 4 about one's own response to *NLMG* giving the lie to the worry that our world is as devoid of thought of rebellion as theirs.
26. I've done so in detail in my papers critiquing Louis Sass: "On approaching schizophrenia via Wittgenstein", in *Philosophical Psychology* 14:4 (2001), 499–514; "Literature as philosophy of psychology: William Faulkner as Wittgensteinian", *Philosophy, Psychology, Psychiatry* 10:2 (2003), 115–124; and "On delusions of sense: A response to J.M. Coetzee and Louis Sass *Philosophy, Psychology, Psychiatry* 10:2 (2003), 135–141 (and in my *Applying Wittgenstein* [London: Continuum Press, 2007]). Cf. also Coetzee's reply to me: "Fictional beings", *Philosophy, Psychology, Psychiatry* 10:2 (2003), 133–4.
27. See p. 143 of Des Pres.
28. p. 183. Cf. on this my exchange with John Foster in the journal *Global Discourse*, http://dx.doi.org/10.1080/23269995.2017.1300442.
29. p. 191.
30. p. 199, emphasis added.
31. And Gollum (as perceived by those who do not know him) is the image of desperate total craving. The image, though not the reality; the reality of Gollum is also the development of trust and community; this is the key plotline of *The Two Towers*. Including the community that comes from acknowledgement of shared plight. The plight of all Dasein that is vulnerable to the attractions of the Ring. (There but for the gaze of Sauron go I. Or as Gollum puts it, "Master [Frodo] cares [for Gollum]. Master knows [what it is like to suffer from the

Ring]." "Once it takes hold of us, it never lets go.") The tragedy of *The Two Towers* is the failure of that project of trust—due, I submit, not to failures of Frodo or of Gollum, but ultimately of Sam (and Faramir). Not even Sam is perfect. (See the beautiful discussion of this on p. 86 of Mark Eddy Smith's Tolkien's Ordinary Virtues: Exploring the Spiritual Themes of The Lord of the Rings (Downers Grove: InterVarsity Press, 2002)).

32. A case in point is Shelob, the great spider, who is of course a metaphor for this feature of fear, too: the more you struggle, the worse you trap yourself in the web (of fear), and thus the more surely you attract the attention of this monstrous thing who will consume you. The more that, scared, you flail around, the worse it gets: Frodo is almost paralyzed by this fear, understandably. The actual paralysis that Shelob induces when she stabs you is merely a literalization of what has probably already occurred in your mind and your heart.
33. And faith's advantage over hope is that hope tends still to be too "internally" related to fear.
34. And many more die when, on Wormtongue's intelligence, the Uruks place explosives in the drain below Helm's Deep, thus destroying one of its defensive walls. (Is the point in part that Wormtongue's free bad deeds are challenging and piercing Theoden's withdrawal strategies, and thus [inadvertently] facilitating Theoden's final wild and successful world-involving stratagem, of going out to meet his deepest fears—to meet what promises to be death—head-on, and vanquishing them, with the aid of suddenly arrived previously alienated kinsmen?)
35. Thus I've been including hobbits, including Gollum's kind, under the heading of "humans."
36. It is important also to bear in mind remarks such as Gandalf's when he says that he pities even Sauron's slaves.
37. When Gandalf comes to speak to Theoden, Grima Wormtongue says to him and to the company, "Why do you lay these troubles on an already troubled mind?" Theoden shows clearly at this point the advantages of depression: not being bothered by your son's death, for instance. Theoden withdraws because it seems safer . . . to give up hope. The hopeless person is "successfully" inoculated against further disappointment. (I return to this point in this book's Conclusion.)
38. It is notable that much the same happened to Aragorn, on the way to Helm's Deep; and to Frodo, in the first film: I return to this way in which the three focal figures of the Fellowship all "resurrect," in n.73. (Add to which that something similar happens to Theoden, when he is saved from the Lord of the Nazgul, albeit briefly, by Eowyn, on the Pelennor Fields. And that he can die proud, because he has answered Gondor's call to stand in solidarity—not given up on them, his fellow "men," other human beings in need.)
39. All Saruman's grandiosity and megalomania before Wormtongue, and on the balcony then before his heaving army, is turned to dust when Merry and Pippin and the Ents appear. We see him running to and fro, pointlessly, in his refuge which is now a prison. He is stuck in the tower that had seemed to be the icon of his power.
40. My presentation hereabouts, re the trumping, under most circumstances, of the warrior virtues by the ordinary virtues, is influenced by Tzetvan Todorov's magnificent book, Facing the Extreme: Moral Life in the Camps (Berkeley: Henry Holt, 1997). See also T. Des Pres's The Survivor: An Anatomy of Life in the Death Camps (Oxford: OUP, 1976), a crucial influence on Todorov's book, and Gita Sereny's Into That Darkness (London: Deutsch, 1974). It is worth noting also that there is at least one moment when we see something straight out of the world of the camps, writ large, in Jackson's film: Frodo's vision, in

Galadriel's mirror, of the future of the Shire, and the future existence of hobbits, should Sauron gain the Ring and triumph. (It is perhaps also worth bearing in mind once again that Tolkien wrote his book from 1936 to 1949.)

41. A threat suspected vividly by Faramir, whose level of trust in his relations with Frodo and Sam runs roughly inversely to Gollum's. The question of trust, of acknowledgement, of shared plight and community, runs throughout *The Two Towers*, then; for of course, as argued above, the questions facing Treebeard, Merry and Pippin, and facing Theoden, Gandalf, Aragorn and the Elves, are, in the end, little different.
42. All page references to the version translated by the Muirs and Sterns, published in Description of a Struggle and The Great Wall of China (London: Secker and Warburg, 1960).
43. pp. 190–91; my emphasis.
44. pp. 192–93, my emphasis.
45. p. 194.
46. An important point implicit here is that theologically standard "faith" is therefore, unfortunately, faithless. A true faith does not involve the fear of God. For God could not be something that one should so fear, and faith could and should "trump" any fear that bubbled up.

    There could (as I shall go on to note) be a spiritual critique here of the whole of supernaturalistic theism. This is a major weakness of Eddy Smith's otherwise good book on The Lord of the Rings, Tolkien's Ordinary Virtues: his chapter on faith is actually a chapter on mere superstition.
47. See for instance the first-person account offered in To Schiz and Back: A Philosophical Memoir of Hell by Jeff Cumberland (Lulu Books, 2011), www.lulu.com/shop/jeff-cumberland/to-schiz-and-back-a-philosophical-memoir-of-hell/ebook/product-17539100.html.
48. Consider, in the *Fellowship of the Ring* film, the way that the Lord of the Nazgul appears to materialize out of thin air, at Weathertop. Almost as if he materializes from Frodo's fear.
49. The thought that God is onto me, onto my badness, might seem very different from the thought that God is simply malevolent, "punishing" me despite my innocence. But the two thoughts are actually very close to one another; if one has the latter thought, one will probably search out the former, as a possible explanation; if one has the former, one will periodically reach for the latter, as a "palliative," as a preferable alternative, a refuge. They are often moments in the same psychological dialectic.
50. Descartes might be defended in the following way: he employs a fixed concentration on the mind as a literary and philosophical device, so as to avoid having to worry about the kinds of terrors that embodied creatures must worry about if their imaginations run riot. But this defence fails, for it fails to take into account that minds without bodies can still be tortured: by [delusions of] cognitive penetration by others, by sheer confusion, or by [real or feared] moral corruption. Furthermore, it fails to register that becoming only a mind can be a torture, in itself. Losing the world, losing one's sense of embodiment, when it happens to people, is often experienced in that deeply painful way.
51. Here, one should think once more of the way in which the Ring is clearly presented in *Lord of the Rings* as having an internal relation to Sauron, and also to all over whom it comes to have power: crucially, Gollum (as manifested especially in Gollum's extraordinary way of speaking) and Frodo. A useful comparison is to Daniel Paul Schreber (see his Memoirs of My Nervous Illness; see also Sass's Paradoxes of Delusion on him): Schreber's God was essentially tied to him. One of his great theological discoveries was that God is not self-standingly omnipotent, but dependent for his power and well-being upon man.

And Schreber's way of speaking, the form of his language, was sometimes quite as peculiar as Gollum's. Intriguingly akin to it, at times, in fact.

52. A striking note of optimism in the tale is that the corruption, with hobbits and men (with the exception of the Nine, who had their own rings), is never complete. This fits with my suggestion above about the "non-existence" of the "muselmann."
53. Here we might usefully think of Hegel and the master-slave dialectic, as a psychologically acute corrective to Descartes's cruder model. Cf. n.68.
54. While one can be tortured worse, if one is still hoping: see on this Villiers de L'isle-Adam's powerful short story, "A torture of hope", *The Strand* 131 (Feb. 1891). See also Vladimir Nabokov's greatest psychological/Modernist novel, Invitation to a Beheading (New York: Putnams, 1959).
55. Something like the "demiurge" reading could be supported by The Silmarillion (London: Harper Collins 1999 [1977]), in which we learn that there is a larger more ancient figure standing behind Sauron: Morgoth. However, I don't draw much on The Silmarillion here, because it lies wholly outside the films.

    My argument in this chapter has the implication then that the concept of God comes or could come both from psychopathology (which is what Gnosticism looks like), and/or from love. The question is, which version of God will we run with / give our allegiance to.
56. p. 174, The Survivor; underlining added.
57. p. 176.
58. p. 176.
59. Faramir thus overcomes rather better than Boromir did the temptation he feels to fall for the Ring. Appearances notwithstanding, Faramir is stronger of character than Boromir. Boromir is more like Denethor, wanting power without having to earn it; doesn't Denethor in fact deliberately choose the son to represent him at the Council of Elrond who he believes more likely to fall for (and to take) the Ring?
60. It will be objected that Sauron was of course not technically omnipotent even with the Ring. Indeed; but this supports my reading. One's fear that one is completing a malign other into omnipotence outstrips the facts, the possibilities, which are always worse and yet better than this: one's responsibilities never end (unless one dies); one is never completely abject. Sauron perhaps fantasized his own omnipotence (". . . rule them all"); but it was a fantasy. Compare here my discussion of Kafka's "burrower" and his fantasies, above.
61. For this enemy seems as concrete as can be. It is even a man, not an Orc or some other monster. We might say (after Wittgenstein) that Faramir's deconstruction of "the enemy" is a grammatical remark.
62. This is in (the full-length version of) the Jackson film only: it is one of many moments, especially in *The Two Towers*, where I find the films—especially the full-length versions, not the first cinematic releases—superior in terms of emotional and philosophical aplomb to the books.
63. A similar exercise would be worth carrying out *vis-à-vis* the men and other creatures who fight for Saruman against the Rohirrim—Saruman's speech as he sends them out to fight suggests that they would be merely revenging themselves for past injustices at the hands of the "horsemen." In a world ruled by kings and swords, why not?
64. Who are simultaneously mere echoes of and "essences" of men. They are our craving.
65. This is the contrast Todorov (in the introductory part of Facing the Extreme) draws between the heroism of the Warsaw ghetto uprising (against impossible odds), as against the unnecessary warrior moment of the Polish uprising against the Nazis in 1944 (in a situation when by contrast there was real hope, and just

waiting would have been better). The intriguing thing about the Warsaw ghetto uprising is that, though virtually none of those who rose up survived, still their survival rate was higher than the survival rate of those who allowed themselves to be taken to the extermination camps.

66. See also p. 362 of "Of the rings of power," in The Silmarillion (London: Harper Collins 1999 [1977]).
67. It is important to note here the precise parallel between what happens to Gollum and what happens to Frodo: Sam (along with Frodo when Frodo is put under duress by Faramir) caused Gollum to lose trust in Sam (and Frodo); now Frodo too loses trust in Sam. All can lose trust, as well as find it.
68. Gollum's mutual connection with Frodo in *The Two Towers* as I presented it above could of course also be profitably read through Hegel's famous account of the "master-slave dialectic." Bear in mind that the slave never loses their subjecthood altogether; far from it. Gollum is trusted by Frodo more than Boromir is; bear in mind that "Gollum possessed the ring for nearly five hundred years, and yet he did not fade as did all nine of the Ring-wraiths, who had begun as mortal Men" (Tolkien's Ordinary Virtues, p. 87).
69. And through the non-linguistic consequences of that language. Take the wonderful scene earlier in *The Two Towers* where Gollum (through a dialogue filmed increasingly as if between two people) succeeds in banishing (N.B.: not integrating) his "bad" self, his selfishly protective ego. In a stroke of genius, we see Gollum at the end of this scene looking slightly frantically around to see if "bad Gollum" has really gone, or if he is still present. . . . This is what it is like. As with the Ring; one concretizes the bad part, one alienates it into being; and then one can't quite believe that things have got better, even when they have.

    Or consider the opening sequence of *The Return of the King*, in which there is a powerful presentation of Gollum as suffering from alienation, (and) from the internalization of the others' image of one. As Gollum then puts it, in beautiful unconscious paradox: "We wept, to be so alone." Knowing that one is one, one nevertheless exists as two.
70. Incidentally, this is one respect among many in which Philip Pullman's masterly body-loving epic death-of-God satire on monotheistic religion, His Dark Materials (Reading: Scholastic, 1995–9), is surely a reworking of Tolkien. Pullman, unlike Tolkien, of course means to criticize Christianity itself.
71. This moment is directly comparable with that I emphasized in the previous chapter, of Kowalski saying to Stone, as he drifts away from her, and as she refuses for a long time to accept this, "Ryan, you're gonna have to learn to let go . . . I wanna hear you say you're gonna make it."
72. Did he perhaps feel broken by his taking the Ring for himself within Mount Doom, as Levi felt broken on comparing himself utterly unfavourably with the "last one" of the Auschwitz rebels? Some wounds, it seems, never heal.
73. There is a close comparison here with Christ's self-overcoming in *The Last Temptation of Christ*. Scorsese's Jesus is especially alike to Frodo's similarly self-sacrificial Christ-figure in having to give up a wonderful ordinary life and in achieving his quest through, as well as in despite of, a psychotic episode (this is the "last temptation"). It is clear that Frodo, Gandalf and Aragorn are for a number of reasons partial-Christ-figures in the story, not least in that they all on at least one occasion fairly literally are resurrected (come back from the dead).
74. Whose paranoia and loneliness is perhaps the worst of all. . . . I develop this point in the latter part of my discussion of *LOTR*.
75. Compare from The Silmarillion: "[T]oo great was the evil power of [the One Ring] for any of the Wise to yield, unless like Curunir [Saruman] he wished himself to become a tyrant and a dark lord in his turn" (p. 364).
76. This is the world that Frodo glimpses in Galadriel's mirror, a possible future.

77. The scare quotes around "concluded" mark that Frodo's psychological quest continues upon his return to the Shire. The difficulty in recovering from quasi-psychotic states, again, is simply that it seems that one's quest can never be truly over, or at least (and this comes to the same thing) that one can never be confident enough that it is, relative to the awfulness of the possibility that it isn't. Thus Frodo cannot return to his old life. See for instance the way in which he "necessarily" withdraws into himself, at Sam's wedding.
78. Pascal's thinking in "The Wager" is, technically, quite compatible with faithlessness—or indeed with schizy belief in a malign demon—as with a truer faith. In citing Pascal positively here, I am perhaps implicitly thinking more of Chomsky's reading of Pascal (see n.172)—a de-divinized reading in which the necessity of social hope for good social outcomes is elaborated—than of Pascal's original text. And I am thinking of the *contrast* with Descartes; for whom it was knowledge, not faith, that was paramount. *The Lord of the Rings* teaches the dangers of a knowingness that would replace faith.
79. Cf. the deeply impressive Phillip K. Dick based works, *Blade Runner* and *Total Recall*, which also combine the sceptical impulse and the interest in psychopathology.
80. Auden says, of "The quest hero" (see p. 45 of his essay of that title, in N. Isaacs and R. Zimbardo, Tolkien and the Critics [Notre Dame: University of Notre Dame, 1968]), "I am conscious of myself as unique—my goal is for me only."
81. Aragorn to Frodo at Amon Hen: "Frodo, I have sworn to protect you." Frodo rightly asks, in reply: "But can you protect me from yourself?"

    This is what Frodo sees, wearing the Ring, on Amon Hen: "At first he could see little. He seemed to be in a world of mist in which there were only shadows: the Ring was upon him. . . . // All the power of the Dark Lord was in motion. . . . Mount Doom was burning, and a great reek rising. Then at last his gaze was held: . . . Barad-dûr, Fortress of Sauron. All hope left him. // And suddenly he felt the Eye. There was an eye in the Dark Tower that did not sleep. He knew that it had become aware of his gaze. A fierce eager will was there. It leaped towards him; almost like a finger he felt it, searching for him. Very soon it would nail him down . . ." (The Fellowship of the Ring [London: Unwin, 1981 (1954), pp. 520–521]). What as a small being you find when you touch the horror of the violence of the world, is also a despair, and a terror. If that despair and terror afflicts you enough, what you find rising up in and at you then in a vicious circle is an overwhelming malign force: and that IS the "Lord" of the Rings (or a Gnostic God).
82. Compare also Gandalf to Frodo, outside Moria: "You must trust yourself, trust your own strength." Hope grows on hope, fear on fear—faith on faith.
83. This suggests a challenge to "Rational Choice Theory" (and more generally to neo-classical economics), which pretends to analyze situations as they are objectively. If every human situation involves faith in the sense just discussed, then there will in the first instance be no human situations where the laws and maxims of Rational Choice Theory are applicable. Alternatively put: Rational Choice Theory is the very disease of which it takes itself to be the cure, for it threatens to dissolve our (much-needed) sense of how what situation we are in is always in part a matter of our orientation towards that situation—our hopes, our degree of faith (in ourselves, in others) etc. If and when Rational Choice Theory is ever true, that is because people have wittingly or unwittingly renounced their freedom to transcend it; and the more that that freedom is renounced, and the more that people notice others behaving as if the theory is simply true, then (unfortunately) the more it tends to become—self-fulfillingly—true, empirically. (For some development of this line of thought, see my "Three strikes against the difference principle," *International Journal of Green Economics*, 5:2 (2011), 167–183).

84. And what one can be reduced to, in the nightmare, is being nothing but craving, for safety, for not having to feel (vulnerable) any more.

 Compare again the account of the desperate states seemingly of utter "animality" (yielding only to utter lassitude)—of mere life—produced for instance in some concentration camps; compare the great analysis of this by Primo Levi (and also perhaps Agamben's writings on the same, and Phil Hutchinson's Shame and Philosophy (Basingstoke: Palgrave, 2008)), as discussed above.

85. Thus, as well as noting a general affinity between Plato on the one hand and Tolkien/Jackson on the other—in the parallelism drawn (systematically, in the Republic) between psyche and polity—we should also note that Tolkien's creation is in important respects even closer specifically to Plato's "Ring of Gyges" (and to some of Wagner's Ring cycle) than meets the eye. It is about being able to kill the king, and/or to become the king, both literally and in fantasy, and it is about how literal kingship/lordship is necessarily infected with the psychology of fantasy lordship, of omnipotence of mind or spirit. It is about wanting to be the lord of the rings, the lord of all, and about how one can indeed seem to be that—though necessarily not for long, in a private world (and in no other).

 For such a "private world" is, I have implied, no world at all—compare the Wittgensteinian work on this of John McDowell and Charles Travis.

86. Aragorn is among those moving beyond the pure warrior ethic. One can see clearly, by the time of the battle before the Black Gate, in the eerie calm before that storm, that that ethic, and the hit (the drug) of fighting, of killing, of power, is starting to be overcome. One can perhaps imagine a happy sequel to the rather pathetic coronation scene that follows: a sequel involving not only the prioritization of non-violent relinquishment over violent possession (a prioritization increasingly explicit in the deviant non-war-like structure of the so-called "War of the Ring") but also the relinquishment of monarchy in favour of something like a (non-fear-based) less hierarchical proto-democracy. The bowing of all before the four heroic and ordinary hobbits, after the coronation, is possibly already a gesture in that direction.

87. As Edmund Fuller says, in "The lord of the hobbits," in N. Isaacs and R. Zimbardo (eds.), Tolkien and the Critics (Notre Dame: University of Notre Dame, 1968): "With so heavy odds, against so formidable an adversary, a significant factor provides one hopeful element in the grim web of Sauron's network of agents, tracking down the Ring. In Sauron's very nature, he is incapable of anticipating the policy adopted by his enemies. He cannot conceive that they would voluntarily relinquish the Ring and destroy it, for it would be incompatible with his nature to do so. Thus, the one move that he does not expect is that they would themselves convey it to his very threshold in an ultimate renunciation and destruction of its power." Similarly W. H. Auden, in his "The quest hero," on p. 57 in the same collection: "[T]he mistakes which Sauron makes to his undoing are the kind of mistakes which Evil, however powerful, cannot help making because it is Evil. His primary weakness is a lack of imagination, for, while Good can imagine what it would be like to be Evil, Evil cannot imagine what it would be like to be Good." Yes and no: *Lord of the Rings* I think actually does enable us to understand—to imagine—what it is like to imagine that (some-)one is Evil, which is all, I submit, that is actually possible, "Evil" being only an "absence," a strategy. The "evil" seek to reject whatever is good within them, as a "rational", powerful (though literally self-defeating) self-defence strategy.

88. And just as Frodo's failure at the last to rid himself of the Ring makes sense in terms of the extraordinary grip of the fantasy of achieving safety through achieving power, so this "accident" that accomplishes what he cannot also makes complete sense in terms of what I have argued is the central "moral"

of the tale: the injunction never to give up on a human being, the endless call to love other precious sentience. As Rose Zimbardo remarks, on p. 102 of her "Moral vision in The Lord of the Rings," in Isaacs and Zimbardo, Tolkien and the Critics, "The hobbits are the common man, who does not seek out the opportunity for great deeds, who prefers his bounded life in The Shire. Yet the hobbits are the heroes. . . . Their peculiar excellence is not heroic honour but love. Frodo is finally saved because he has pitied Smeagol". Compare also Marion Zimmer Bradley's not-dissimilar and very subtle interpretation of Gollum's "accidental" completion of the quest, on p. 123 of her "Men, halflings and hero-worship," from the same volume.

89. If we compare Frodo and Aragorn, paired protagonists in *Lord of the Rings*, at the end of the tale, they present a seemingly stark contrast: the one is in the end unable to savour victory, unable to live in the ordinary paradise of the Shire; the other is utterly able to—but at what future cost? On my reading of *Lord of the Rings*, the unsatisfactory political solution will lead to a return of the repressed—we might simply say, of the Ring. The psychological will come to haunt the political. A Book 4 of Tolkien's saga might then be "The corruption of the King," "The madness of King Aragorn," or even "The return of the Lord [of the Rings]."
90. What happens during Gollum's final fall, as depicted in Jackson's film, brings to mind the way in which Kafka's burrower wants to contemplate the safety of his burrow without entering it. In contemplation, it is always better than in reality. Thus Gollum in his last moments does not put on the Ring: he merely gazes lovingly at it, captivated.
91. In *World Fellowship of Buddhists Review* XLI:4 and XLII:1 (Oct. 2004–Mar. 2005), 14–22.
92. Loy and Goodhew also claim (p. 15) that the suffering that Frodo experiences on his way to Mount Doom makes him "stronger and more compassionate." This seems to me only half-true; as I have argued above, the psychological subtlety and tragedy of Tolkien's presentation is that Frodo is also in a way destroyed, terminally alienated, by the pathologies he has had to undergo. His was an ungainsayable sacrifice.
93. p. 243 of Humphrey Carpenter's The Letters of J.R.R. Tolkien (Boston: Houghton Mifflin, 1981).
94. On p. 21.
95. This is a larger part of the explicit topic of *Avatar*; see the coda of this chapter.
96. Quoted at p. 14 of David Loy, The World Is Made of Stories.
97. I suggest what this could look like here: https://medium.com/@GreenRupertRead/religion-after-the-death-of-god-the-rise-of-pantheism-and-the-return-to-the-source-54453788bbaa.
98. Aragorn loves Boromir at the latter's dying, because he is human, with all the weakness and beauty and strength and trying of us, encapsulated.
99. Or better, as D. Loy and L. Goodhew put it, on p. 16 of their "The karma of the rings: A myth for modern Buddhism," *World Fellowship of Buddhists Review* XLI:4 and XLII:1 (Oct. 2004–Mar. 2005), 14–22: "So is Frodo's journey a spiritual quest or a struggle to help the world? In The Lord of the Rings, these two are the same. . . . Middle-earth needs to be saved, not denied or escaped. The goal is not another world but another way of living in this one, even as nirvana is not another place but a liberated way of experiencing this one."
100. This theme of sacrifice (and of resurrection from the death that the main three protagonists of *Lord of the Rings* all experience) reminds one keenly of the sense common among many afflicted with schizophrenia that they are a sacrifice. That they are undergoing this for all humanity. Compare for instance Schreber's persistent thinking along these lines. (And I am not asserting that

this sense is wholly missguided.) Recall Frodo's words to Sam, at their heart-rending last parting: "[W]hen things are in danger: some one has to give them up, lose them, so that others may keep them" (The Return of the King [London: Unwin, 1955 (1981), p. 376]).

101. The fantasy being that however far you retreat, you could still retreat further. (Thus at Helm's Deep there was at the end talk even of somehow retreating from the keep, back inside the very mountain.) And in a way, this is phenomenologically true, as Laing's brilliant early accounts of psychosis testify (see e.g. his discussion of "the false-self system," in The Divided Self). But you are becoming (as Bilbo puts it) thinner; shrouded in darkness, alienated even from yourself. You would cease to be, or you will become mere life, and neither is what you hoped the strategy would achieve!

102. It is also arguably a classist, sexist and prejudiced-against-the-disabled-and-misshapen text. My answer to the last of these three not-unreasonable charges is implicit in my answer to the racism charge. Briefly, my response to the first two charges is this:

Yes, *Lord of the Rings* is classist, pro-autocracy, pro-hierarchy etc. However, as indicated above, this, its worst blind spot, is "caught up in" the very logic I am attributing to the text: i.e. the tendency towards autocracy and rigid hierarchy (and thus, indirectly, towards classism) is implicit in the logic of paranoia, the logic of the very futile search for safety that *The Lord of the Rings* successfully deconstructs and overcomes. The classism and monarchism of the text will prove a final victim of a successful following through of my reading. One must extend the central insight of the text to overcome the vestiges in it of the very pathologies whose shape and nature it throws into relief.

As to the sexism charge: This seems to me mostly misplaced, especially (though not only) with regard to the films. Galadriel is as utterly splendid an overarching heroine as one could hope for; Arwen's saving of Frodo from the wraiths and from wraithdom, and (more crucially) her becoming tied later to the fate of the Ring—her daring to hope for the future of Middle-earth—are key aspects/moments of the text as I read it; Eowyn's becoming a liberal feminist icon as she cross-dresses her way to war and out of her "cage," and her "Macduff" moment as she saves Theoden from the Lord of the Nazgul; these are not mere accidents of the text, they are pivotal to it, all three.

103. An indication of the non-existence of the Orcs and of Sauron's other creatures and vassals, of their character as mere extensions of or toys of Sauron, who in turn is our fantasy, is the way in which they instantly lose all direction, and are mostly just swallowed up by the Earth, at the Ring's dissolution and Sauron's dissolution. Real creatures, soldiers, would not have to stop fighting just because their distant commander was struck down. One critic who has more or less understood this is Hugh Keenan, in his "The appeal of The Lord of the Rings: A struggle for life," in Isaacs and Zimbardo, Tolkien and the Critics: "In The Lord of the Rings, Sauron can be . . . viewed as the objectification of the fears and self-destruction (death instinct) of the inhabitants of Middle Earth."

104. When for instance the Orcs—extensions of Sauron, who is in turn a projection of Frodo (i.e. of you and me)—fight the ghost-knights (who are the demons of/projections of Aragorn), then we have, one might say, a battle fought entirely within the mind. (Cf. also n.103.)

105. See the epigraph to this book.

106. Paranoia increases in proportion to power.

107. The Ring is really nothing—just a (charming) piece of old metal. One thinks it is something—one's craving tells one so. But that's really only the craving. (This can be felt or seen for oneself, in meditation.) The craving substantializes things that are nothing, but for one thinking them something. Compare the following

remark of Wittgenstein's, from Culture and Value, p. 13 (Oxford: Blackwell; revised edition, 1998): "Compare the solution of philosophical problems with the fairy tale gift that seems magical in the enchanted castle and if it is looked at in daylight is nothing but an ordinary bit of iron."

108. Again, why after all *does* Sauron want the Ring? Is it not for the same reason that we all do?
109. For political power that is not entirely consensual, genuinely democratic, as power sometimes is within Quaker communities, and within non-violent direct action "affinity groups."
110. The kind of power dreamt of for instance by "vanguardists," as with Lenin and his successors, in the East, in Tolkien's time.
111. By the same token, it would be a grave mistake to read this chapter as predicated upon an othering of the "mentally ill," the "psychotic." I hope rather to have made perspicuous how the drive towards psychosis is entirely natural under certain circumstances, and further, even that a human being who felt no temptation to madness would barely be human at all. "Mental illness" is a constitutive or necessary possibility of and within rationality. And the idea that the "mentally ill" are deeply other than "us" is itself a pathological idea, one that "they" and "we" are both vulnerable to.
112. Compare Lévi-Strauss's introductory remarks to The Raw and the Cooked: Introduction to a Science of Mythology (London: Pimlico, 1995 (1964)).
113. See Chapter 4 of David Nicholson-Lord's Green Cities—And Why We Need Them (London: NEF, 2003) for discussion.
114. Recall here Nietzsche's remarkable argument, "If there were a God, how could I bear not to be a God? Therefore, there are no Gods!"
115. Deep thanks for help (with this work on *LOTR*) to Phil Hutchinson. Thanks also to Patrick Curry, Alun Davies, Peter Kramer, Frances Dunlop, Preston King, Emma Bell, Chris Cowley, Louis Sass, Alison Roberts, Graham Read, Juliette Harkin, John Heaton, and audiences in the Film Dept. Seminar at UEA, at the Public Lectures on Mental Health at UEA, and at a Film and Philosophy Conference at the University of Liverpool.

***

116. See the start of the "full" script for Avatar (www.imsdb.com/scripts/Avatar.html): The first words of the script are spoken in a voiceover by Jake, accompanied by images of what will turn out to be Pandora: "When I was lying there in the V.A. hospital, with a big hole blown through the middle of my life, I started having these dreams of flying. . . . Sooner or later though, you always have to wake up . . ."
117. Her organs, as we had it in Chapter 3.
118. Of course, an ex-Marine, a fighter, someone valued not just for his DNA but for being a special kind of soldier, is not strictly speaking an "everyman." But I think the elision is justified: for this is everyman as everyone (or: many; or at least: many American men) wants at least sometimes, in patriotic or adventurous mood, to think of themselves, or to identify with. He is an everyman, an ordinary guy, in that he is rank and file, one of a mass who doesn't think for himself, a grunt, a "jarhead" as he calls himself—until he starts to awaken.
119. Tsu'tey also calls the dreamwalkers "demons." "A demon in a false body." An intriguing possible parallel is highlighted here with Sauron, or the Balrog and other "demons" in *Lord of the Rings*.
120. I mean the word "feel" emotionally/metaphorically, here. 3-D isn't yet virtual reality. But in *Avatar*, seeing literally is believing, and you are asked to feel what you see and what you believe. So the metaphor is not an empty one.

121. See e.g. https://en.wikipedia.org/wiki/Pandoran_biosphere#Na'vi.
122. Cf. the ecopsychological work of Mary-Jayne Rust et al.
123. This is from p. 25 of Chris Johnstone's "The Avatar effect," *Permaculture Magazine* 64 (2010), 25–26. (Johnstone is a colleague of Rust's (see n.122)).
124. A charge issued by some critics and "Leftists," but made most famously by the then Pope. Thankfully, my guess would be that the new Pope, co-author of possibly the most significant ecological document of this generation, the Laudato Si, would take a different view.
125. Is the film a tale of simply going "back to nature"? No, in this very important respect: What we are offered in the film, on Pandora, is an indigenous culture, a culture which offers us a big change: taking ourselves to be part of the ecosystem, again. The lesson of *Avatar* (and *Apocalypto*) for us can't simply be: go back to the jungle. Because there are already people living there.
126. This is my belief; I cannot justify it here, but readers interested in following it up or as yet unconvinced may wish to consult Georgescu-Roegen's work. (See also www.edie.net/news/news_story.asp?id=18222&channel=0&title=UN+uses+World+Environment+Day+to+outline+economic+case+for+restoring+ecosystems)
127. The kind of direction that we need to be moving in instead is indicated here: www.triplepundit.com/2010/05/sustainability-as-usual-isnt-good-enough/. Cf. also the vital contribution of "permaculture."
128. The first tentative opening of Jake towards the transformation that he needs to undergo, around this time, is perhaps why the seeds of the sacred tree touch him. He is just starting to tune in to what they represent and protect.
129. The "etc." here is important. As Joshua Clover puts it, on p. 6 of "The struggle for space," *Film Quarterly* 63:3 (2010), 6–7: "The blue anthro-feline Na'vi, three meters tall, are not so much any indigenous people, but rather any number thereof: Native Americans in their natural harmony, or the Urarina of the Amazon rainforest—but no more these than, say, Iraqi natives (cued bluntly by the phrase 'shock and awe')."
130. See for instance this important project: www.sehn.org/Volume_14-2.html. And my own "Guardians of Britain's future generations", www.opendemocracy.net/ourkingdom/rupert-read/guardians-of-britains-future-generations.
131. For example, Michael Peterson et al, "I see you?", http://flowtv.org/2010/02/i-see-you-gender-and-disability-in-avatarmichael-peterson-laurie-beth-clark-and-lisa-nakamura/.
132. One might say, in *Avatar*, all of us are disabled, because we can't fly. But we are all of us potentially en-abled, by being able to fly on creatures that we enter into a relatively harmonious one-on-one relationship with.
133. It is also important to note that, on the journey to Pandora at the opening of the film, Jake is un-disabled, by weightlessness. In space, no one can see that you are "crippled." In practice, you aren't.
134. And, by analogy, in a possible future in which we have poisoned our own atmosphere.
135. Similarly, at the end of the film, as I discuss, she kisses his (human) body as he is about to attempt to transfer permanently to his avatar.
136. PI "Part II" section iv—cf. discussion earlier in this chapter.
137. See e.g. http://james-camerons-avatar.wikia.com/wiki/Pandora.
138. Of course, this point in itself is not new. See for example *A Man Called Horse*, *Dances with Wolves*, *The Last of the Mohicans*. What is new, I would claim, is the involvement of the audience, "therapeutically," transformatively.
139. The phrase famously used by Adorno and Horkheimer early in The Dialectic of Enlightenment to describe the Enlightenment in its actual consequences.
140. For critiques of "sustainable development," see Helena Norberg-Hodge, Ancient futures (Sutton: Rider Press, 2000 (1991)), Debal Deb, Beyond developmentality

(London: Routledge, 2009), and my report on "Post-growth common sense": www.greenhousethinktank.org/uploads/4/8/3/2/48324387/post_growth_commonsense_inside.pdf.

141. Starting, of course, with Jake.
142. Thus the argument of Annalee Newitz's "When will white people stop making movies like *Avatar*?" (http://io9.com/5422666/when-will-white-people-stop-making-movies-like-avatar) completely misses the point.
143. Think here for example of James Frazer, as critiqued by Wittgenstein in his "Remarks on Frazer's Golden Bough." See also my There Is No Such Thing as a Social Science (Aldershot: Ashgate, 2008), co-authored with Wes Sharrock and Phil Hutchinson.
144. Though we should note that, prior to the action of the film, Grace herself apparently had to go through a transformative journey of her own—the early attitude of the anthropologists towards the Na'vi, during the "backstory," was it seems rather more one of teaching them our ways, and trying to get them to wear Western clothes etc. They have already gone through a process of reversing their assumptions qua anthropologists.
145. My There Is No Such Thing as a Social Science (co-authored with Hutchinson and Sharrock) explores at some length how such enlightened anthropology prescinds altogether from the category of science, and how "social science" is a recipe for endless failure to understand the other (and to avoid changing oneself).
146. See e.g. Stuart Kirsch's influential book of that name (Stanford: Stanford University Press, 2006).
147. The term "unobtainium" does in fact have a genuine use in engineering—it refers to whatever one wants but cannot get that would make what one wants to do (as an engineering task) but cannot do possible. See e.g. http://en.wikipedia.org/wiki/Unobtainium. To use it as a name for an actual mineral is a kind of general literalization of this dream or fantasy. To actually obtain unobtainium is as impossible as (say) floating mountains are. This point about the actual use of the term therefore strongly supports the claim I make in the text, and connects also with my discussions of the powerful relation of this film to dreams. There are things it is harmful to dream of—such as actually obtaining "unobtainum."
148. One of the exciting things about the blockbuster films that this chapter concerns is that they both feature on-screen subtitles (*LOTR*'s being for Elvish). Subtitles—who would have thought it? If you had predicted a generation ago that the biggest-grossing trilogy of films and most successful individual film ever made would have prominently featured subtitles as one heard on-screen radically foreign languages, you would have been laughed at as an arthouse fantasist. (Obviously, one might also mention *Apocalypto* here, made entirely in a foreign language—and, quite obviously, encouraging the viewer to travel in a similar direction, ultimately, as *Avatar*.)
149. Evening Thoughts: Reflecting on Earth as Sacred Community, edited by Mary Evelyn Tucker (San Francisco: Sierra Club, 2006), p. 17.
150. In this way, as a struggle of recognition/acknowledgement, *Avatar* is closely connected with some of the other major philosophical movies of modern times, such as *Blade Runner*, *Fight Club*, *District 9*, and the films discussed in Chapters 1–3 of this book.
151. In an impressive forthcoming paper entitled "Look at the shiny shiny! Narrative deficiencies and visual pleasures in Avatar."
152. Even the final visual of the film, as the credits roll, is a point-of-view shot of flying through the sky of Pandora and descending into the canopy of the forest. This is in so many powerful ways a biophilic film.
153. Compare also my discussion of the confrontation of machine vs. avatar/Na'vi, supra to n.160.

154. P. 6 of his "The struggle for space".
155. In this connection, the task of the protagonist in *Avatar* is identical to Deckard's in *Blade Runner*—see Mulhall's writing thereon. That task is also very close to that which Iris Murdoch called upon us to undertake: a task of loving-attention.
156. Just as in *District 9*, which in this connection again demands comparison with *Avatar*.
157. See especially Zettel (Oxford: Blackwell, 1967), p. 382.
158. Here I am thinking of Thomas Berry's "great work," and of Wittgenstein's remark (in Culture and Value) that work in philosophy, like work in architecture, is really work on oneself. I am thinking also of the kind of intertwined micro- and macrocosmic change that (I argued above) *LOTR* concerns.
159. As Nietzsche put it: "[B]ecome the person you are" (The Gay Science 270 (London: Vintage, 1974 (1882)).
160. Particularly striking about the robot-warrior-suits is that they don't have any heads. The head, the intelligence, must be supplied by a human. Sadly, such intelligence is mostly lacking in the colonizers headed by Quaritch on Pandora.
161. A kind of Pandoran rival to (Scott's and) Cameron's Aliens.
162. As Quaritch sees it: "If there is a Hell, you might wanna go there for some R and R after a tour on Pandora. Out there beyond that fence, every living thing that crawls, flies or squats in the mud wants to kill you." This is classic nature-hatred; what it gets right is that, without a "social model" of how to live in and cope with a natural world, without a willingness to listen to it and adapt to it and respect it as the Na'vi have done, it cannot but seem hostile.
163. Justification of this claim that we have an idolatry of science and technology, and that seeing technology as "neutral" is dangerous, can be found in Heidegger's The Question Concerning Technology, and in my own work on the philosophy of science. This is of course not to rail against all technology: there remains a vast role for science and technology in improving our lives, in preventing disaster (think of climate science), and indeed in making films like *Avatar*. But a healthy, non-scientistic relationship with science and technology, giving up the fantasy of inevitable "progress," is some way from where we currently are. (For clues towards it, a valuable text is Joel Kovel's The Enemy of Nature.)
164. The Chinese authorities, in their fear, interpreted the film more or less correctly, like some of the right-wing critics of the film. They perceived its genuinely radical content and potential, and were scared that it could ignite resistance to land grabs etc.; see Itamar Zohar's "China bans *Avatar* from 1,600 cinemas due to fear of popular revolt", www.haaretz.com/1.5049164. (The film might have particularly moved audiences, because of the intriguing visual resemblance between the landscape of the Impossible Mountains and the landscape of China, especially that landscape as imagined/rendered in the classical traditions of Chinese nature art.)
165. See e.g. www.telegraph.co.uk/news/picturegalleries/worldnews/7222508/Palestinians-dressed-as-the-Navi-from-the-film-Avatar-stage-a-protest-against-Israels-separation-barrier.html and www.google.co.uk/images?hl=&q=avatar+palestine+protest&rlz=1B3GGGL_en-GBGB353GB354&um=1&ie=UTF-8&source=univ&ei=y1sKTKDcKY380wSon7Fm&sa=X&oi=image_result_group&ct=title&resnum=1&ved=0CA8QsAQwAA.
166. See "*Avatar's* Na'vi in London to stop Vedanta mine", www.survivalinternational.org/news/6273.
167. See www.radicalanthropologygroup.org/new/Home.html.
168. Think once more of how the atmosphere is apparently poisonous to humans; and how Pandora brings the worst out in human nature at first.

169. The use of scare quotes is advised: the very concept of "natural resources" is a piece of unspeak that aims to make the exploitation of the world easier. As Heidegger has pointed out, treating the world as a "standing reserve" for the use of humankind is a deadly error. A "saving power" needs to arise to counter this, which is why Jay Michaelson puts the pantheistic (or perhaps panentheistic) cosmology of the Na'vi forth, as an alternative to such an outmoded way of thinking: "The sky god tells us that we humans are masters of the Earth; thus, we, like the humans in Avatar treat Earth as a resource to be exploited." (See his "The meaning of Avatar: everything is God," *The Huffington Post*, 22 Dec. 2009.)
170. See e.g. David Ehrenfeld's The Arrogance of Humanism (Oxford: OUP, 1978).
171. As I made clear in discussing *LOTR*.
172. Especially as riffed on by Chomsky: see e.g. p. 355 of an interview, collected in David Barsamian (ed.), Chronicles of Dissent (Stirling, Scotland: AK Press, 1992).
173. This is not to say that there is no such evidence (historical, neurological, evolutionary etc.) of fundamental human goodness! Rather, it is to say that such evidence is always "imponderable," never decisive, often countered or undercut. Something more is needed to undergird collective action and self-confidence. It is also to say that, even if there were no such evidence, then such faith would remain/become our *only* hope of salvation, our only way not to ensure self-destruction through fatalism, inaction, pessimism and consequent self-destructive behaviour. Finally, it is to reiterate that we can make some things possible that seem impossible, that we can create our own future. That the results of the "miracles" (or "fairy tales") that *Avatar* depicts (or tells) can perhaps be made real, given enough human willpower, determination, love and faith. If "our head creates our world," we must dream and think better.
174. As explicated by Morgaine in the Prologue to The Mists of Avalon (Reading: Sphere, 1982), p. ix: "For this is the great secret . . .: that by what men think, we create the world around us, daily new."
175. In this respect once more it rhymes with the "boom-boom" climax of a similarly deep transformative and therapeutic film, *District 9*. It is shocking to find how much one wants the protagonist in that film to kill the South African soldiers. But I think that the *deus ex eywa* that alone brings success in *Avatar* takes everything a stage further than *District 9* by realizing that there is no military solution to problems such as these. We have, rather, truly to win "hearts and minds," in part, through films such as these. Additionally, *District 9* differs crucially from *Avatar* in that the former ends with our protagonist, Wikus, still desperately wanting to become human again, while the latter ends, contrariwise, with our protagonist completing the transformation away from being human. Both have opened to truly seeing the other: thus by the end of *District 9* the "prawns" have become persons to us. But the transition away from human-centrism is more complete in *Avatar*.
176. This movement is similar to the central, brilliant conceit of Justin Leiber's philosophical novel, Beyond Rejection: that the way to start to feel truly at home in a body not one's own is to learn to hate one's original body and way of living and what it stood for.
177. As some reviewers, predictably, have claimed that it is: see especially Todd McCarthy's Review in *Variety* (10 Dec. 2009), which explicitly bemoans the ending as less believable than it "should" and could have been.
178. The title of James Lovelock's book, The revenge of Gaia: why the Earth is fighting back and how we can still save humanity (London: Penguin, 2006), in seeming to suggest otherwise, brings out the danger inherent in the anthropomorphizing

move that has unfortunately tended to be involved in Lovelock's presentation of his concept of Gaia.

179. Deep ecological thinkers such as Aldo Leopold, Susan Flader, John Seed, Joanna Macy, Pat Fleming and Arne Naess have taught us to "think like a mountain." But the framing is unperspicuous (because we don't tend to think of mountains as other than vast rock-like objects). I'll substitute the following way of putting their point: we need to think like an ecosystem. We need to redesign human societies so that they work as ecosystems do, without having any waste; this is the central principle of permaculture. And we need to learn to think like ecosystems, to ensure that we are not living in a way incompatible with ecosystemic resilience among those ecosystems that are primarily or almost entirely outside of human life, too. The Na'vi—and, obviously, real indigenous peoples—can help with both tasks. (See also Helena Norberg-Hodge's Ancient Futures.)
180. This reading is I think confirmed by the song that accompanies the closing credits, "I See You" (sung by a mixed-race woman, Leona Lewis): www.directlyrics.com/leona-lewis-i-see-you-lyrics.html. The song is no great work of art, but the lyrics are quite powerful and interesting, especially perhaps, in the present context, the repeated refrain, "I see me through your eyes" as well as, "I see you."
181. If one searches the Internet a little, one can find many cases in which viewers are speaking of *Avatar* as having changed their lives. I have found evidence of people who have given up their 4x4s, people who have become involved in activism, and many people who speak of having had their eyes opened.

     When I give talks on *Avatar*, I find the audience reactions interesting in this respect, too. For example, I had an ROTC recruit speak of feeling guilt and questioning herself being part of the army.
182. My sense of the hero's journey is strongly influenced by Christopher Vogler's book The Writer's Journey (Ann Arbor: Sheridan, 2007 (1998)). But it also comes from Homer—by way of Tolkien/Jackson, Kubrick, Cuarón. . . .
183. Thanks to my Avatarian co-conspirators Vincent Gaine and Peter Kramer for very helpful comments on an earlier draft of this coda. Thanks also to Chris Knight (especially), to Jeremy Thres, to Tom Greaves, and to an audience at the University of East Anglia. Biggest thanks finally to Naomi Marghaleet for extraordinary and creative help in editing this chapter. For a brief shining moment, we saw each other. . . .

Conclusion
What Have We Learnt?

> What is film criticism?: The process of tracing out the effects of a film in writing that seeks to prolong and increase them.
>
> —Chris Fujiwara

A (broadly Cavellian) thought that has motivated this book is the following: we go to the cinema, ultimately, so as to *reconnect* with the world (and with each other), a world that alienates us from it and from each other (for instance, through capitalism—and the ideology of individualism).

I have offered in this book an *engaged* philosophical engagement with films: not only an interpretation, as in most philosophical readings of films—though "radical" interpretations this book certainly has offered—but (more than that) a way of thinking (and *feeling*) about, *with* and *through* films, a way that I hope is genuinely transformative, ethically, politically, existentially. I have sought to bring out what is already present in these films and, I claim, somewhere in the responses of many viewers to them—*and* to work to reduce your resistance to these presencings, so as to enable the films' effects to be prolonged and increased.

For philosophy, on "our" Wittgensteinian method, mainly consists in changing one's way of seeing and in overcoming one's resistance to such change. This book has explored what that means *in relation to films*. Philosophy, for Wittgenstein, is essentially freedom: not the freedom of selfishness, nor of mere license, but a genuine freedom from intellectual and appetitive enslavement.

The films I've looked at are, we might venture to say, designed or destined to midwife a personal or inter-personal, philosophical and even political *aware-ing*, by means of which we no longer take ourselves to be superior to or alienated from the rest of life. This process of becoming aware, conscious, is a true freedom (from dogmas in which we are drowning, from the heteronomy that rules us), including a freedom from the hegemonic fantasies of freedom itself, fantasies that are killing us: consumeristic, individualistic, "progressive"[1] fantasies, that rule our world.

***

Wittgenstein and Cavell have of course been central figures for the growth in interest in film as philosophy, especially since the publication of my *Film as Philosophy: Essays on Cinema After Wittgenstein and Cavell*, over a decade ago. In *this* book, I have laid out how, after Wittgenstein, "film as philosophy as liberation" can be equally as fruitful for popular films as for "arthouse" films. Of course, in doing so, I trod in the footsteps of Cavell (with old Hollywood) and Mulhall (with modern Hollywood), but the films included in this book have run the gamut from the ultimate "difficult" art-house movies (of Resnais) all the way to *the* highest grossing films of all time. The films considered in Chapters 5 and (especially) 6 are if anything even more popular/mass-market than those that in most cases Cavell and Mulhall considered.

The films which I chose to discuss in this book address, as I see them, important features of our time in its heart of darkness, in a manner *that essentially includes the viewer*. They do not (in the main) lecture or didacticize; they facilitate an open-ended experience of growing wisdom, and of ethical and indeed political engagement.

They have in common *titles* that help make this experience available. In the case of every one of my 12 chosen films, I have dwelt on the film's title, and argued for its appositeness and depth.

I hope to have shown moreover how it is possible to have (and to write) a very *personal response* to a film while simultaneously making would-be *objective value judgements* about it. It is clear, at the same time, that the experience of the viewer will vary to some extent, in part dependent upon the "subject-position" one has (e.g. one's degree of personal experience of melancholia; or perhaps one's gender), in part dependent upon one's spiritual or existential starting point, and so on. This is a feature of Wittgensteinian/"therapeutic" (or, as I increasingly think of it, *liberatory*) philosophy: the resolution of the apparent conflict between the personal nature of an account and the "objective" claims made for it is *via* a proper understanding of how I, following Wittgenstein, take the nature of *philosophy* to be, where the term "philosophical" doesn't come down on the side of objectivity as against the personal. Rather, it *bridges the gap* between the two. This kind of personal response/involvement/continuation is demanded, as we saw, by films such as *Melancholia*, where the viewer is joined with others in an authentic-making "dance," just like Justine painfully becomes authentic. Such "continuation" is also demanded, more self-evidently, by films like *Avatar*, where (like in *2001: A Space Odyssey* before it), the film closes with a direct "invitation" to join in with the awakening that the film has midwifed, exemplified and celebrated. Such personal (or corporate) involvement or continuation is demanded, in one way or another, by *every* film in this book.

In order to be able to genuinely appreciate the quality of—to *see*—such films (as *Apocalypto*, *Avatar*, *Lord of the Rings*, *Gravity* etc.), one has to overcome a prejudice: the prejudice that mass entertainment cannot be of significant intellectual import. (Luckily, a broadly Wittgensteinian sensibility, after the later Baker, places one well to overcome intellectual prejudices: this is in fact Wittgenstein's central task.) I have sought to overcome such prejudice, mainly by "doing" rather than merely saying. I have not spent time discussing Wittgenstein (or Cavell or Mulhall), but rather I have focused primarily on "reading" the films, on suggesting an orientation to them, and on capturing aspects of one's experience of them. This too, I hope, manifests a Wittgensteinian sensibility: by refraining from "theory," I have manifested primarily a set of "examples" designed to challenge and overcome certain specific and problematic intellectual prejudices.

***

In rounding out this book, I'll return briefly to recapitulate and *link together* some of those examples. By dwelling on what connects the 12 films I have focused upon. Most obviously, their common interest in trauma. This interest is present even in the one film of the 12 focal to this book where it is perhaps not obvious: in *2001: A space odyssey,* I argued for its presence in HAL's experience. In nine of the films, there is also a strong interest in recovery, figured as re-emergence, as an awakening to and *as* real freedom. (In *LYiM*, *NLMG* and (I argued) *Solaris*, that interest is absent . . . but *profoundly* absent; present by way of its absence.)

Consider for instance the way that in Chapter 1 I brought out a way in which *Waltz With Bashir* (and *Apocalypto*) midwifed the overcoming of certain self-serving prejudices, by first encouraging one viscerally to inhabit them, and to generate a sense of righteous complacency, which then gets pulled out from under one. If, as I have urged, we desperately need to *wake up*, then we got a good "picture" of *how*, from these films. The radically "therapeutic" (in Wittgenstein's sense of that word) force of *Waltz With Bashir*—which, as it follows the protagonist (who is also the filmmaker) in his psychotherapeutic journey, facilitates one's painful emergence from a prejudice in favour of the ("Western") Israelis, lies in its ability subtly to *involve* the viewer in a distinct lack of acknowledgement of the other. This dissociation persists until the end of the film, when suddenly, through real, full-colour footage, we *see* the survivors of the Sabra and Shatila massacre. Suddenly, that is, the reality which was skilfully ignored or suppressed erupts out and the protagonist (and the viewer with him) finally understands what he has participated in, and what he repressed.

My analysis of *Waltz* explained what is otherwise mysterious: *why* this film is an animation. The animation (and the music) serve the purpose of showing the viewer the disturbing distance Israeli soldiers put between themselves and the Palestinians/Arabs. The very texture of the film tends towards compromising the viewer, and dehumanizing the victims:[2] until the very end, where Sabra and Shatila finally appear, in documentary reality. Suddenly, the humanity which had been skilfully/wilfully suppressed/ignored comes out, and the protagonist (and the viewer with him) finally understands what he has participated in, and what he repressed. The victims, who were merely drawings, and whose voice one never heard prior to that point, suddenly explode into one's consciousness with full power and reality.

Similarly, at the close of *Apocalypto*, our sense of complacent righteous identification with the victims is swept away and we are left with something much more challenging and unsettling, as we realize that we are framed as structurally similar to the perpetrators. We are forced then to think again about our place in the world—and about how to preserve the place(s) of indigenous peoples, and jungles/rainforests and their creatures, who share this Earth and might yet be able to remind us how to live lightly on it.

This kind of process is potentially painful. The pain is necessary. It is part of what it is to become free and to re-embed oneself in reality.

Our situation is alarmingly grave, partly because we have allowed ourselves to "float free" of that ecological reality.[3] I argued in Chapter 3 that *Never Let Me Go* is literally almost true, and that it *is* metaphorically true, and prefigures possible futures in which that metaphorical truth would be starkly visible, disaster triumphant, the dread result of our "Enlightenment," our "progress," our "growth."

The really alarming thing about *Never Let Me Go* then, even if I am right that, strictly speaking, its scenario is conceptually impossible (in a way—that I set out—complementary to *The Road*'s), is how very close to being true it still is—and furthermore how tempted we are to take the lazy route of taking it to be completely realistic (and thus having a great "excuse" for giving up). Are the "students" in the story really so different from us? For we too, nearly all of us, spend most of our lives entirely caught up in pathetic minutiae and in non-rebellion, while close analogues of the *NLMG* scenario play out, and the world burns (threatening to produce within the lifetime of our children scenarios alarmingly close to that of *The Road*).

The real encouragement however that I argued that this bleak film yields is that there are those who can face up to this reality. The "those" who can see the world of *Never Let Me Go* as it is are: us. The "audience."

What do I see the audience of these films being led to undergo, and potentially to realize? Much of this book has consisted of a kind of deeply sympathetic film "criticism" operating at (I hope) a high but nevertheless accessible level of philosophical sophistication. Paradoxically, this sophistication has been in the service mainly of a kind of learning that is in the end very simple. I have not, I hope, been making too many clever academic points, but rather, "reading" these works and finding a re-focusing on and re-telling of vital "timeless" wisdom *for our time*. A wisdom found in what Buddhism calls the possibility of enlightenment, in becoming liberated from thought-forms that bind us within hegemonic common-nonsense, and in rising to realize that freedom, amid a context of what needs to be ecological rejuvenation and restoration.

Consider for instance the way in which so many of the films I've written about help us to think and feel appropriately (about thinking and feeling appropriately) about grief and loss. Vital in so many ways—but especially, surely, at a time when we are already having to grieve for what the living planet has lost as a result of our actions.

Recall in this context *Hiroshima Mon Amour*. As we saw in Chapter 2, this masterpiece of Alain Resnais's (the script compellingly authored by Marguerite Duras) opens with a long, deeply strange, "alienating" sequence in which a female voice seeks to offer *models* for how one can (she could) understand what happened at Hiroshima in 1945, based on her visit there, her time at the museum, by watching reconstructions of the events of those days etc.; and in which a male voice denies, quasi-sceptically, that she actually has succeeded in gaining any such understanding. His voice says what appear to be patently false things, such as "You saw nothing in Hiroshima." The sequence demands, I suggested, to be heard as a philosophical dialogue, in which one voice denies that what would seem ordinary understanding is available because the events in question are so mind-boggling, so extraordinary, that ordinary modes and models of access do not work. And this is important because if we do not understand the history of Hiroshima it will, as the other voice tells us, happen again.

The film as a whole should, I suggested, be understood as an attempt to facilitate for viewers the possibility of gaining some understanding of the epochal, terrible calamity and crime of Hiroshima, through a route of indirection. Forced to acknowledge that one does not understand such an event through normal routes, in such a case, one comes to the possibility of an understanding, of some perspicuity, indirectly, via some kind of understanding of extreme, appalling happenings on a *micro*-scale. Considering how it is possible therapeutically to work through calamitous grief, one comes to see how it might conceivably be possible to understand (and thus finally to work through) the same kind of phenomenon on a ghastly macro-scale.

Hiroshima Mon Amour could be viewed then as providing a quasi-Wittgensteinian "argument" against pseudo-scientific modelling and in favour rather of *objects of comparison* (cf. *PI* 130–2, explored explicitly in Chapter 3), in relation to cases in which modelling would crudify, reduce and offer a false, cheap *illusion* of knowledge. Cases such as understanding a vast, unprecedented act of destruction as occurred at Hiroshima; and cases such as are present in philosophy, which by definition are extraordinary. And I set out in Chapter 2 how a key—in my view a virtually conclusive indication—that this kind of "reading" of the film is what it requires, is provided by the final 20–25 minutes of the film, when it becomes clear that it is inadequate to regard the film's two lead "characters" as "*characters*" at all: they are more like metaphors for the cities/countries from which they come, which are (or should be) engaged in a process of mutual understanding, a difficult process, one whose difficulty cannot be underestimated, but which the film might help to midwife.

Last Year in Marienbad pursues an examination of trauma and trappedness in the past, I argued, on the level of psychosis, as *Hiroshima Mon*

Amour does it, explicitly, on the level of neurosis. (There is less explicitness in *Marienbad* precisely because there is less "insight" in psychosis. Neurosis can know itself as psychosis cannot.) *Hiroshima* considers the phenomenon of neurosis on both macro and micro levels; it begins with the traumatic macro (the insanity of a world posed to potentially "nuke" itself, as first demonstrated possible at Hiroshima), takes one through the traumatic micro, and (as I argued in Chapter 2) takes one back then to the macro, in order to apply the lesson. We really might have a better chance at avoiding the madness of nuclear holocaust if we follow something like this trajectory, and refuse to forget the horror of Hiroshima.

Melancholia does something similar, moving with blatant but also graceful clarity from the macro to the micro, and facilitating in us potentially a grief-in-advance for the life on Earth we are destroying, a grief that might yet help us to avert such destruction. *Melancholia*'s focus is on neurosis (and grief), but one inevitably thinks of the world-catastrophe that it depicts both as a probably soon actually-to-exist result of collective psychosis, and possibly as the world-catastrophe *experience* typical of extreme "rational" psychosis, of the kind depicted by Louis Sass in the closing chapter of his book *Madness and Modernism*. *Solaris*, by what is on balance a contrast, while of course intensely interested in neurosis and grief, ends, intriguingly (like *Marienbad*), in apparent psychosis. *Gravity*'s interest in and working through grief (and depression) is, as I have pointed up in Chapter 5, clear; it could be read in conjunction with *Solaris* as providing a more hopeful take on the same basic phenomenon of loss.[4] All five films, I would argue, work then essentially by investigating at a micro level what is (then) to be "applied" at a macro level. In *Melancholia* and in *Hiroshima*, this is starkly clear, with the move back and forth from the personal to the political/planetary. In *Gravity* and *Marienbad*, it works more subtly; I showed how one ought to see the trajectory of (Ryan) Stone/of *Gravity* as a metonym or allegory for the ecological challenge we face, for the gravity of our situation; and how *Marienbad* ought to strike us eventually as an account of the malady that is ultimately responsible for that challenge and more.

If one is stuck, then one needs to find ways of freeing oneself up. If we are in Plato's cave, we need to turn around to actually see the world that makes the moving images.[5] As I've periodically noted in this book, to get lost in the films is the obverse of what we need: that's as it were what happens to the two main protagonists of *LYiM*, and to all the astronauts in *Solaris*. Rather, the role of filmic attention, in the "ecology" of the present work at least, is to function as what *LOTR* sees rightly as a *temporary*, tactical or meditative retreat, a deep shared attention that then enables us to *see* and to go forth together.

Hunters in the Snow is literally a picture of coming home. Of the return leg of journeying—that was our central preoccupation in Chapters 5 and 6. But in *Solaris*, like in *LYiM*, there is no actual coming home, only a fantasy of it. *Solaris* ends with what *appears* to be the successful, emotionally resolving accomplishment of the hero's journey (back to Earth, and back to the family home). But while Justine, at the death, achieves equanimity and loving kindness, when we see the camera pan back from Kelvin and his father embracing, we see that Kelvin is still on Solaris. Stuck, in his own mind.

Solaris is explicitly a meditation on what it is to be human (in part through the various characters, including 'Hari' herself, considering her status), and on whether we know ourselves. As if to underscore the point, a bust of Socrates inhabits the library where perhaps the most important scenes of the film take place. Wittgenstein inherits and modifies the ancient Socratic task of philosophy: self-knowledge, of the right kind. We can (and need to) come to know ourselves better; and this *essentially includes* seeing ourselves as involved with one another, rather than staying put in a "private" realm. We can come to understand grief better, to know madness better, to see relations of these with philosophy. Doing so *essentially involves* seeing (e.g.) the difference between grief and depression—and the similarity between epistemology and madness.

The pay-off of doing so is not only potential benefits in terms of understanding and gaining sympathy towards the suffering involved in psychopathologies of others or of ourselves. It is the same process *vis-à-vis* (macro) psychopathologies of our time, of our civilization. And thus some greater autonomy may be realized with regard to these; an ability to no long feel compelled by them, and to find alternatives. A better prospect of regaining the kind of wisdom explicitly at issue in *Lord of the Rings*, *Avatar*, *Apocalypto*, *The Road*, and implicitly in *2001*, *Gravity*, *Hiroshima Mon Amour*, *NLMG* etc. The kind of wisdom which would mean that we would actually refuse to let each other go, refuse to let the living Earth die. And in particular, as I argued in Chapter 5, that we will refuse to be distracted by the wildest of fantasies[6], about fleeing Earth:[7] these fantasies are catastrophic at a time when we are wreaking catastrophe upon the Earth, because they allow us to half-think that maybe that wreaking isn't so catastrophic after all, because maybe Earth is just a staging post for us anyway. (The corollary of my thought here is that, ironically, only once we become clear that we have to stay and get our own house in order would it become morally kosher to think (let alone plan) seriously about whether there is any way to live in space/beyond the Earth. Otherwise, there is too much "moral hazard" involved in the enterprise.)

Finally then, reconsider briefly the way in which I have sought to read (view) *Lord of the Rings* as ploughing much the same furrow—at one and the same time eco-politically and psychopathologically, macrocosmically and microcosmically—through my interpretation of this "hero's journey" that rather radically rewrites our sense of what kind of person a hero ought to be (as does *Gravity*). In Peter Jackson's films of *The Lord of the Rings*, we *see* on the screen—we *experience* viscerally the point of view manifest in—a pathological search for safety, for surety. (Most notably, we see this vividly in scenes in which one reaches for the Ring, for invisibility, for escape to a private realm that is one's own, a realm where one can be lord and master.)

As I set out in Chapter 6, one can feel watched, in *Lord of the Rings*, by *the Ring*, by a mere piece of metal, a symbol. A closely related, wonderful inspiration of the films is the giving of a *voice* to the Ring. One comes to notice the indistinct, disturbing, seductive voice *of the Ring*, as one watches with attention. (We might usefully compare the phenomenon of voice-hearing as encountered on psychedelic drugs, as well of course as some voice-hearing in "mental illness.")

Again and again in *Lord of the Rings*, the suggestion becomes that any effective refuge through withdrawal, refuge of the kind most strikingly sought by Frodo, by Theoden, and by Denethor, can be at best temporary. The physical structure of Minas Tirith largely echoes that of Helm's Deep, and, in *The Return of the King*, the effect of Denethor's paranoid and immunizing-himself-against-disappointment-through-assuming-and-bringing-about-the-worst retreat within it (and within himself), to a tomb of self-imposed death and denial, is punctured only by the sending for riders (riders who have ridden *out* from Helm's Deep, and *out* now, indeed, from Rohan) that Gandalf has managed to engineer, meantime. Denethor's retreat (which is pretty much terminal for him) in the third film mirrors and deepens Theoden's retreat (which *he* manages to reverse, with Gandalf's aid) in the second. Again, we saw this in detail in Denethor's refusal even to see that Faramir is alive, his cursing Rohan for having betrayed him when he didn't even try to call for Rohan's help and yet the help is nevertheless arriving; and so on.

In Chapter 6, I viewed key aspects such as these of *Lord of the Rings* as an allegorizing of this desire to retreat, to avoid having to come out and meet one's demons. An allegory of the desire for absolute power—which is in the end only attainable within one's own mind. I thus gave a philosophical rendition of the *Lord of the Rings* as a subtler and nastier moral threat than Descartes's demon, and thus for seeing Jackson/Tolkien as offering a philosophical corrective to Descartes, filling in the gaps in his presentation of what it would actually mean to imagine a malign demon of infinite or (better) of very great power. The really disturbing, the more deeply psychologically challenging notion (than Descartes's), the clear and distinct idea that can unworld one, is that *that "malignity" is quite incomplete without us*, without our existentially ongoing participation. The desire for the Ring is the desire to BE the Lord of the Rings, to become invulnerable through being "all-powerful"; complementarily, the desire to give the Ring to Sauron *is* the desire to already be abject before such an "all-powerful" Lord of the Rings; both are (pathological) efforts to escape from the ordinary lived human condition of "limited" always-already-embodied existence, the worst fear of which is being confronted, not with a malign omnipotent demon, but with a malign demon who can only be completed *by you*.[8]

For what can be (or rather: feel) worse than having lost all hope? Answer: having a very slight hope only, that you are continually depressed about, fearing your inadequacy even to try for it.[9] For when you still have a slight hope *you can still get disappointed*. Or: you can still *fail*. Alternatively put: there is something worse than the negative sublime . . . namely, one's own role in the ongoing creation of this negative sublime. The infinitely awful is not as bad as one's completion of the infinitely (or almost infinitely) awful.

The genuine hope that *this* in turn yields, of course, is that, in being taken to this worst place of all, this place worse even than total powerlessness—i.e. the place where you are potentially to blame for the worst possible outcome—you have of necessity taken back a little agency. *And that's all you need*. As soon as you know that you have some possibility of resistance, that despair is not total, that faith has not been eliminated and may yet be effective, then you can recommence the struggle, and eventually bring the elixir home. And, as we saw at the end of Chapters 3, 4 and 5—and especially at the end of Chapter 6, in discussing the unusual way in which *Avatar* seems to end the hero's journey prematurely—that is the ultimate "pay-off" of these films: placing *us* the viewers in a situation where the cinema is no longer about the world viewed, but about the world-to-be-changed.

And this was a further strand in the defence of my interpretation of *The Lord of the Rings* as a struggle to overcome the temptation to see life as a struggle between Good and Evil, with this "meta-struggle" being ultimately a psychological/philosophical struggle to understand that "evil" is only lack, and to give up the desperate self-defeating longing for safe refuge from evil and power over it. Serious "mental illness" *often feels to the sufferer like a titanic battle between good and evil*. Just as *Lord of the Rings* often feels that way. A really good film about madness must be a film that doesn't seem to be obviously and definitely about madness (cf. *Last Year in Marienbad*), even though the threat of madness is somehow subtly yet powerfully present to it. (For even true fear of "madness" fears most of all that one is actually sane, that this is really happening.)

And this points strongly beyond simplistic versions of the warrior ethic that tend to dominate popular understanding of what *The Lord of the Rings* is about. It is not about fighting and winning and defeating the enemy, *not even in the battles*. (And towards the close of Chapter 6, I suggested that the same holds of *Avatar*.)

That my interpretation of the story in terms of the *development* of a soundly non-aggressive yet non-withdrawing cast of mind—*neither fight nor flight*—is fruitful and may even be simply correct can, I think, be seen through the way in which as the story goes on the protagonists learn better and better to ride out to battle when it is needful, and care less and less about conventional victory.

Let's review the sequence once more, briefly:

- Contrast the withdrawal within Helm's Deep and the failure to reach out for support beyond Rohan with the wonderful (if desperate) riding out from the keep of Helm's Deep. And contrast in turn the latter with the freely chosen decision of Theoden's and of the Rohirrim to go out to support the people ("men") of Minas Tirith, on the Pelennor Fields.

- Similarly, contrast the unnecessary and quite pointless assault on Osgiliath led by Faramir on Denethor's insistence (there is not *necessarily* anything good about "riding out," in the world of *Lord of the Rings*) with the freely chosen, dignified and courageous decision of the warriors to go out, to charge forth and willingly sacrifice themselves, one and all, "For Frodo," before the Black Gate. (They wait a long time, watchfully, while they are gradually surrounded, spinning their "diversion" out for as long as possible. They are not precipitate. When there can be no more waiting, in Jackson's film, then they astonish by once more going *out* to the enemy, rather than quaking; thus they symbolize and actualize their psychological victory over dread and despair—which brings its reward when, despite their certain loss, they win, because they have after all given Frodo enough time to dispose of the Ring.)[10]

A final overarching suggestion implicit in Chapter 6 then might be said to be this (perhaps-surprising) one: that *The Lord of the Rings* is a *Modernist* text, thus bearing closer comparison to some of the films considered in earlier chapters than one would have *prima facie* expected. Or rather: that the antagonism between Modernism and "Ancientism" or "Mediaevalism" (and it might *seem* far more natural to class *Lord of the Rings* as having an ancient or mediaeval psycho-political sensibility than a Modern one) is usually overstated (e.g. by Louis Sass). Or again, rather (and this is the most important—psycho-philosophical—point, that can if necessary let interpretation of Tolkien's/Jackson's text go altogether): that an Ancient vision of the world as inhabited by spirits and the Modern worldview (at least, according to Kafka, Weber, de Chirico, Orwell, Foucault, Sass etc.) of oneself as watched/haunted by scrutinizing agencies and eyes and by voices—by vision, and by language—are almost two sides of the same coin. On the Ancient vision, each thing has its spirit. On the Modern version, each thing has it eye/voice. Both Ancient and Modern visions spring from our quest for meaning and our tendency to fear, and for fear to go out of control. Both are "rational." But both constitutively risk prescinding from healthy normality.

***

I will not seek to sum up further what I have learnt in writing this book, let alone guess/surmise what *you* may have learnt from reading it. In particular, I hope you have not received the impression that my "readings" (viewings, writings) have unearthed what I regard as decisively or definitively *the* most important aspects of these films. I have aimed to highlight certain key aspects, for sure; and, typically, neglected aspects. I have sought forcefully to render these salient, well supported, sufficient, and to thread them together. But: this book concerns how films can facilitate a true freedom. Freedom from the intellectual hegemony (of humanism, "progress," "growth," "individualism", scientism, technophilia etc.) that is currently killing us.[11] It would be a performative contradiction to insist that you sign up to a new hegemony.

On the one hand, then, we need a new philosophy, which this book seeks to elicit, that is suitable for our time: a shared ecological philosophy, a way of being in the world / on this Earth that can last. An alternative to the crude consumeristic freedom that enslaves us. "But" on the other hand, the new-old philosophy that we need needs to include constitutively a true freedom. A freedom of mind and heart, that will manifest both in the radical non-conformist kinds of action that will be needed if we are to have any chance of saving ourselves (perhaps the kind of action seen at times in the latter portion of *Avatar*, for instance), *and* in a willingness to *keep on* thinking, and not to leap simply from slavery to one hegemon to slavery to another. If we do not keep on thinking, we will fall into the dreadful kind of complacent, narrow-minded, techno-scientific "progressivism" of (for instance) Christopher Nolan's *Interstellar*, and we will at best leap from one frying pan to another. Rather than, as *Gravity* I think encourages us to, putting out the fires raging here, putting our own house in order. Drawing on the way that *2001* is a space *odyssey* in that it is all about coming home. All about the tether back to Earth. The "gravitational pull" of life on (and to) Mother Earth. A pull that, I claimed in Chapter 5, is perceived by both Bowman and HAL alike. Compare *2001* with the proposal of *Gravity*: by their end, both films appear to be far more about celebrating and preserving life on Earth than they are about the wonders of technology and space travel. Both films *call upon us to awaken*. To actually become aware of the absolute wonder of what we already have, that is so desperately at risk. To awaken to life *on* and return *to* Earth, not to move pointlessly with ever increasing rapidity (around and) beyond it.

Part of why we tend to be so asleep is that there is a *crisis of attention* in our society. Our attention is constantly grabbed, rarely deep, and thus we often are in little position to perceive what the films in this book are seeking to wake us up to.

As I hinted in the Introduction to this book, the marvellous thing about film is that the very experience of sitting down to watch a film, especially in a cinema, can help here. Prolonged quasi-meditative shared attention itself creates "*space*." It creates a space where there is some real freedom (for reflection; for thought, or feeling; for the groundwork of significant change to be done; maybe for something *beyond* thought), and slowness.

The spirit of *Interstellar*, and of many mainstream films (especially most romcoms, war films, superhero films, thrillers and horror films), however, contradicts that sense of spaciousness—instead, we endlessly hurry onto the next thing (the next high, the next product or gadget, the next planet, the next special effect). Many movies are just a kind of incarnated ADHD.

But the kind of films in this book oppose that sense of rush, and can grow that sense of spaciousness. Even when they do not seem to, such as in

Apocalypto's rush of Jaguar Paw's escape and hunt, or in the swashbuckling warring portions of *Lord of the Rings*, I think they ultimately point in the same direction. For, as I suggested in Chapter 1, the rush and the hunt in *Apocalypto* set up the final reveal, in which more radically we come to recast who the hunters and the hunted are (i.e. the "Mayans" will turn out to be the ones who *we*, unjustly, render virtually extinct; it is *our* empire that is, ultimately, on trial here); while I argued that the battles in *Lord of the Rings* ought ultimately to be interpreted psychologically, not literalistically.

We might therefore call the films examined in this book *contemplative*.[12]

If we can only slow down, and be *present*, then we will experience a new freedom, and we will be far less prone to devastate our ecology.[13] The encouraging thing about cinema is that it tends to encourage such slowness and presence in its very medium, in its very setting. And while many mainstream films run away from that possibility into an MTV-style or video-game-style aesthetic, the encouraging thing in particular about the films found in this book is that they tend to encourage such presence, and a concomitant reflectivity. This is so of these films as a whole, and necessarily so; they need to be watched, sometimes at least, in a sitting, and preferably in the dark, with other people whose phones are off (and preferably in a cinema). But the possibility of presence by way of attention can be appreciated most intensely perhaps in certain scenes in them. Recall the languid self-conscious splendour of the whole "overture" to *Melancholia*. Or the majesty of the "levitation" sequence initiated by focusing on *Hunters in the Snow*, in *Solaris*. Or the slow viewings of Earth from space, and of teardrops, in *Gravity*. Or the immersive "therapy" memory sequence in *Hiroshima Mon Amour*. Or the sequences of Jake starting to see the world and to be able to learn, from Neytiri, in *Avatar*. Or the apparently glacial movement of the "rebirth" sequence that closes *2001* (apparently glacial; even though the whole sequence actually only takes a couple of minutes).

If we are to stop hurtling towards destruction, one of the most effective ways to do so may surely be: to stop. To attend awhile to something that encourages instead contemplation, especially of the alternatives to endless hurtling and endless destruction.

I hope that you find the ways I have encouraged one to see the films looked at in this book not only productive of a set of new perspectives, nor only even facilitative of learning, but actively freeing. Freeing from our culture's cultural blinkers, as well as from dubious assumptions concerning the allegedly inevitable limitations on the liberating powers of mass-market/commercial films. Freeing from received "wisdom" about the particular films analysed here: yielding freedom to see them in a new way, be that mine or yours. And freeing to put these films and the sensibilities they yield "to work" in the active struggle to save freedom, fairness and the future itself from the vortex which our current trajectory threatens to throw us into.

Many today in the film-studies world are unwilling to make judgements of quality about the films they study. This book has not followed that pattern.

This book was intended to both reflect and further the filmic midwifing of a process of change and engagement in the reader-viewer. It would thus have been pointless to spend time on films which lack the capacity intelligibly and intelligently to help create such change. Moreover, this book was not intended primarily as me saying this or that about films that happen to interest me, though obviously, and at times centrally, there has been something personal about the views expounded here. The book is rather intended primarily to make available what the films themselves "say." Again, it is pointless to spend time in such a context talking about films that say nothing, or that say only banalities, or that say interesting or even deep things badly. It is only to the point to spend time principally on films that "say" deep things, subtly or movingly. And, further still, as I have emphasized throughout, the scare quotes are essential: such "saying" cannot be merely fully paraphrasable content: if it is, then the film is still not really liberating. Not actually doing philosophy and engaging the reader-viewer in philosophical activity in the sense in which, following Wittgenstein, I want to speak of true philosophy.

No: Film as philosophy as in the best and widest sense of the word "therapy," as loving-freedom, demands to be written in a different spirit. This book had to be focused primarily on those films which are actually good and deep enough to be worthy of such an attempted mode of responsiveness to them. (That has been a vital criterion of selection of which films get discussed herein.)

One of the nice things that emerges from the examination I've undertaken is the homages to great quality works of cinema by possibly equally great later works. We've seen how *Melancholia* depends on and reworks and "opposes" *Solaris* and *Last Year in Marienbad*. How *Marienbad* directly complements and "completes" the picture of *Hiroshima Mon Amour*. And how *Gravity*—and *Avatar*—both mirror, rework and end on similar notes to *2001: A Space Odyssey*.

The way that a work ends is vital to assessing its quality. If all the threads draw together effectively, then that is very different and far more satisfactory than if they do not. And, if they do, then there is an opportunity for "aware-ing."

Often, an important film will end with a kind of a "twist." I do not mean the kind of merely clever twist that ends a film like *12 Monkeys* or *The Usual Suspects*. I am referring to the kind of far deeper forcing of reflection that ends a film like all those in the book but especially perhaps *Waltz With Bashir*, *Apocalypto*, *Hiroshima Mon Amour*, *L'année dernière à Marienbad*, *Melancholia*, and *2001: A Space Odyssey*—or indeed *The Wicker Man*,[14] *Memento*,[15] *Nocturnal Animals*,[16] *Persona*, *White God*, *Monsters*[17] or *Dogville*. Such films disclose something, or facilitate (a seeking after) wisdom, and part of their satisfactoriness in doing so is the way that they leave one pressed to keep doing so, after the credits roll. They are something we can learn from; often what we learn is something very old, yet also something very much for our time.

We appear to have learnt very little as yet from the profound and probably fatal destruction we are wreaking on our civilization by way of the profound and heart-rending destruction we are wreaking on our planetary home.[18] Paying loving attention to the films that are featured in this book (and to other films that I could have analysed along similar lines, including *Children of Men* and *Manhunter*,[19] *District 9*, *The Hunger Games* "trilogy," *Son of Saul*,[20] *Thin Red Line*, *That Obscure Object of Desire*, *WALL-E*, *Koyaanisqatsi*, *Persona*[21] and *Fight Club*) could be one key element of, or at least a prelude to, the kind of active loving attention to each other and our Earth that is needed if we are to have any chance of transforming our civilization in the way required and at the speed now required. It may well be too late to save this civilization. But it certainly will be, if we fail to learn from the fate writ large in some of these films and averted in others. (And, even if this civilization *is* doomed, a genuinely ecological successor-civilization that could and should rise from its rubble would surely resonate with the kind of insights available from works such as *Avatar*, *Apocalypto*, *The Road* and more.)

I hope you have enjoyed and learnt something from the film-philosophy contained in these pages. I commend these films to you. And I look forward perhaps to you doing your bit to move forward from and through them.

Here's to a manifestation of a freedom worth having.

***

A final thought. I have emphasized in this Conclusion the way that a number of films considered in this book work beautifully in one classic way that films seem naturally to work: by *showing* us a microcosm, and tacitly (or openly) encouraging us to "scale that up" to a macrocosm. But I have also noted the way that some of the films considered here complicate that picture. For example:

- *Hiroshima Mon Amour*, as we have seen, effects a huge movement from the macro to the micro and then, in a strikingly deliberate (because "alienating") way, back to the macro again towards the end.
- *Melancholia* considers depression and then equanimity as a rational response to inevitable mortality, potentially on a planetary scale. But it also takes a planet (one vastly bigger than our own, even) as a metaphor for depression. So here we flow in both scalar directions, somewhat seamlessly, as the film weaves its way through our individual and societal neuroses and their potentially/actually dire consequences.
- The battle scenes in *Lord of the Rings* I suggested one experience as macro "models" of mental states. And I suggested that the mental, the psychological (the psychopathological) in *LOTR* in turn models the political. So, in *LOTR*, we move at times from (seeming) macro to micro and then back out to (properly) macro again.

This brings out something important. Most strikingly in Chapters 4 and 6, but also in every single chapter in the book, there is not only an investigation, *involving* we the "audience," of what it would mean to be free in thought, what it would mean to have an enlightened society, and/or what it would be to achieve ecological sanity, but also—and as should already be clear from what is implicit in those very terms or concepts—a potential contribution to our understanding and achievement *of mental health*. If at times in this book I have emphasized the importance of the macrocosm, that is because I think it liable to be neglected in a time of rampant individualistic ideology (and narcissism), and simply because, unless we get the macrocosm sorted, there probably won't *be* (m)any microcosms left in a century or less. But the microcosm is just as essential—and, in many of the films I have considered, actually just as focal. Properly understood, the two need to be combined, inextricably. A film-philosophy of enlightenment and ecology concerns our awareness, our minds, our communities, our common future *without distinction*.

I think that a way of summing up what we find when we find these films might then be this: we heal ourselves by healing the world, but we also heal the world by healing ourselves. To neglect either is to neglect both. *They are not separate*. Healing is necessarily the re-finding of this *wholeness*.

Every one of the 12 films I have explored with you in this book makes its most important contribution of all by enabling a possible liberation through an *experience*: such as the kinds of consciousness-shift that come, to an open mind, at the end of *Waltz*, *Apocalypto*, *Hiroshima*, *Melancholia*, *2001*, *Gravity* and *Avatar*; or the kinds of perspective by way of "impossible" nightmare that are afforded variously by *Never Let Me Go*, *The Road*, *Marienbad*, *Solaris* and *Lord of the Rings*. If we experience those experiences together, so very much the better; and yet they must be our ownmost. You cannot outsource philosophical work to another.

The films focal to this book offer in some cases routes towards justice, including (in some cases) with regard to non-human beings. They offer as a whole a set of routes towards ecological health. *Thoroughly interlinked* is their manifesting of a possibility of enlightenment. Of psychological or spiritual growth. Their offering of or even facilitating emergence from mental ill-health. And that is something much needed at the present time—and much sought after. So perhaps the kind of wisdom that I've been seeking with you, through these films, can truly prove popular.

And, in the darkness of this time, that really is an encouraging thought.[22]

Notes

1. For my critique of the very widespread fantasy of "progressivism," see my "Wittgenstein and the illusion of 'progress': On real politics and real philosophy in a world of technocracy," *Royal Institute of Philosophy Supplement* 78 (Jul. 2016), 265–84, www.researchgate.net/publication/305451081_Wittgenstein_and_the_Illusion_of_%27Progress%27_On_Real_Politics_and_Real_Philosophy_in_a_World_of_Technocracy.

2. Facilitating eventually one's—painful—emergence from a lazy prejudice (in favour of the "Western" Israelis). The painfulness here might be profitably compared to the painfulness of Jake's too-slow journey to decency and being an ally of indigenity, in *Avatar*.

   This relates to the profound connection I touched on earlier between *Waltz* and *Avatar*: both are directed towards the mindset of those whose minds need changing ("us," basically), not towards the mindset of those who are already mentally liberated (respectively: Arabs etc. in *Waltz*, indigenous peoples etc. in *Avatar*).
3. See http://kevinanderson.info/blog/avoiding-dangerous-climate-change-demands-de-growth-strategies-from-wealthier-nations/, www.thelondoneconomic.com/opinion/climate-change-once-we-no-longer-deny-it-then-we-just-might-have-the-will-to-try-drastically-to-change-course/14/03/ and www.truthandpower.com/rupert-read-some-thoughts-on-civilisational-succession/, if further buttressing be needed for this thought.
4. It is intriguing to note that George Clooney, who plays the character lost in the present-time action of *Gravity*, remade *Solaris*—and that the film's producer was James Cameron (director of *Avatar*).
5. The comparison between Plato's cave and the cinema is an obvious one; it has been nicely developed by Nancy Bauer in her essay in my Film as Philosophy collection.
6. I mean this term in the sense that Iris Murdoch or Simone Weil employ it; fantasy in this sense is morally wrong, a mere escapism.
7. See this piece by John Michael Greer, for why it makes so little sense to contemplate living beyond Earth: https://worldnewstrust.com/the-terror-of-deep-time-john-michael-greer.
8. We might say: in the Hell mood, God and the Devil are one. Sauron is the God of a "Godless" world, the Devil as God, as worse than—more terrifying than—God (*because* you are not entirely abject before Him).

   Though the truth—though it is only a truth acceptable and accessible to a happy consciousness—is that, if God were looking at the soul, the heart, the mind, the life of one agonized by themselves, experiencing life as a tormenting problem, facing great guilt, then She *would feel great compassion*. That is: any God worthy of the name looking at an isolated being, a being lost in "Ring-world," would experience what the Buddhists call "karuna," the great compassion. Watching one feel watched by the Eye of Sauron, one can feel little else but this gently enveloping loving sympathy.
9. And here is an answer to the old mystery of subalternity, of why the oppressed submit to their oppression. Because a fully defensive, defeated posture, a giving up of hope, is less painful than the maintenance of a small hope.
10. The riding out from the keep at Helm's Deep and the journey to the Black Gate manifest the kind of character praised by Todorov (in Facing the Extreme) in those who defended the Warsaw ghetto by attacking the Germans when the latter tried to liquidate it. By contrast, the pointless assault on Osgiliath insisted upon by Denethor is more like the "splendid" but unnecessary and militarily pointless Polish-led Warsaw uprising as Todorov reads it. One can conduct oneself with dignity even when victory is impossible. In the mythic world of *Lord of the Rings*, such dignified conduct is rewarded, whereas pointless waste is not.

    So, in roughly Todorovian terms: Theoden, Aragorn, the elves et al. are learning the ways of dignity and care and courage as manifested by "heroes," by warriors; Frodo and Sam are learning and manifesting the ways of dignity and care and courage as manifested by "ordinary folk." "The ordinary virtues" in the end trump "the heroic virtues," *even for warriors*. This is how there is something of a revolutionizing, here, of the sense in which, in the present work, I speak of "the hero's journey": the ultimate heroes are ordinary people. Such as the implied viewer: it is they—we—you—who must act, if the future is to be saved.

11. See my "Gaia is dead: We have killed her, you and I", https://medium.com/@GreenRupertRead/gaia-is-dead-we-have-killed-her-you-and-i-bb040b1d1fff; my "A case for genuine hope in the face of climate disaster", www.thelondoneconomic.com/opinion/a-case-for-genuine-hope-in-the-face-of-climate-disaster/09/03/ and my "Climate change is a white swan", https://medium.com/@GreenRupertRead/climate-change-is-a-white-swan-52ae656f5ba1.
12. I am thinking here of D. Z. Phillips's Wittgensteinian-Rheesian conception of philosophy, as found in his books Philosophy's Cool Place (Ithaca: Cornell, 1999) and Religion and the Hermeneutics of Contemplation (Cambridge: CUP, 2001). (I think that there are problems from a Wittgensteinian point of view with this conception, as compared to a resolute/liberatory conception, but those scholarly problems need not concern us, in the present work.)
13. I expand on this point, by introducing the idea of a "thrutopia," an attempt to return us to living in the present as a way through our crisis of the present and future, here, in my "Thrutopia: why neither dystopias nor utopias are enough to get us through the climate crisis": www.huffingtonpost.co.uk/rupert-read/thrutopia-why-neither-dys_b_18372090.html.
14. I'm referring to the original version: see my exploration of the importance of its mode of ending, in "The new *Total Recall*, the old *Wicker Man*": http://thinkingfilmcollective.blogspot.com/2013/11/the-new-total-recall-old-wicker-man.html.
15. See Hutchinson and Read, "*Memento*: A philosophical investigation," in Film as Philosophy.
16. My thought being that the way *Nocturnal Animals* ends, in flatness, absence and *disappointment*, including probably initially on the part of the viewer, then forces a reassessment—of one's own warped desire for drama, in a film (or perhaps in life).
17. See Phil Hutchinson's "With power to frame the world comes great responsibility: Gareth Edwards's *Monsters*", http://thinkingfilmcollective.blogspot.com/2013/10/with-power-to-frame-world-comes-great.html.
18. See my "This civilization is finished": http://greentalk.org.uk/this-civilisation-is-finished/. What have we done? Condemned our civilization to end. What have we learnt? Not enough—yet. What is to be done? Learn something of a different order—awaken—and so transform ourselves and our civilization or its successor.
19. See my "Popular films as philosophy" (http://thinkingfilmcollective.blogspot.com/2013/10/popular-films-as-philosophy_19.html) for a sketch of the approach I would take to these two films.
20. *Son of Saul* is a masterpiece of point of view in the sense explored in this book.
21. See my take on this masterpiece of Bergman's, and on *Fight Club* as a kind of interpretation of it, in the final pages of my "Wittgenstein as unreliable author / unreliable narrator," in Philosophy in the Condition of Modernism.
22. Thanks to Naomi Marghaleet for editorial assistance.

Index

Page numbers in italic indicate a figure on the corresponding page

For Product Safety Concerns and Information please contact our EU
representative GPSR@taylorandfrancis.com
Taylor & Francis Verlag GmbH, Kaufingerstraße 24, 80331 München, Germany

www.ingramcontent.com/pod-product-compliance
Ingram Content Group UK Ltd.
Pitfield, Milton Keynes, MK11 3LW, UK
UKHW022000180926
475701UK00003B/122
9780367664916